Effortless Cloud-Native App Development Using Skaffold

Simplify the development and deployment of cloud-native Spring Boot applications on Kubernetes with Skaffold

Ashish Choudhary

BIRMINGHAM—MUMBAI

Effortless Cloud-Native App Development Using Skaffold

Group Product Manager: Alok Dhuri
Publishing Product Manager: Shweta Bairoliya
Senior Editor: Nitee Shetty
Content Development Editor: Tiksha Lad
Technical Editor: Maran Fernandes
Copy Editor: Safis Editing
Project Coordinator: Deeksha Thakkar
Proofreader: Safis Editing
Indexer: Subalakshmi Govindhan
Production Designer: Ponraj Dhandapani

First published: September 2021

Production reference: 1090921

Published by Packt Publishing Ltd.
Livery Place
35 Livery Street
Birmingham
B3 2PB, UK.

ISBN 978-1-80107-711-8

www.packt.com

Contributors

About the author

Ashish Choudhary is a software engineer and has over 10 years of experience in the IT industry. He has experience in designing, developing, and deploying web applications. His technical expertise includes Java, Spring Boot, Docker, Kubernetes, IMDG, distributed systems, microservices, DevOps, the cloud, and the general software development life cycle. He is an active blogger and technical writer. He has delivered talks to some renowned conferences such as GitHub Satellite India and Fosdem. He is also a strong advocate of open source technologies. He has been contributing to various open source projects for quite some time. Ashish believes in continuous learning and knowledge sharing.

This book would not have been completed without the patience and motivation of my lovely wife, Shefali, and my little son Ayansh.

About the reviewer

Sirinat Paphatsirinatthi is the director and co-founder of KubeOps Skills. He is interested in cloud-native technology and DevOps culture building for organizations using SRE recommended practices from Google. Currently, he is a speaker disseminating knowledge to the top leading financial services in Thailand and the Kubernetes/Docker Thailand User Group. He is also the community leader of Cloud Native Bangkok.

Table of Contents

Section 2: Getting Started with Skaffold

4

Understanding Skaffold's Features and Architecture

5

Installing Skaffold and Demystifying Its Pipeline Stages

6
Working with Skaffold Container Image Builders and Deployers

Section 3:
Building and Deploying Cloud-Native Spring Boot Applications with Skaffold

7
Building and Deploying a Spring Boot Application with the Cloud Code Plugin

8

Deploying a Spring Boot Application to the Google Kubernetes Engine Using Skaffold

9

Creating a Production-Ready CI/CD Pipeline with Skaffold

10
Exploring Skaffold Alternatives, Best Practices, and Pitfalls

Other Books You May Enjoy

Index

Preface

Tooling around Kubernetes has changed drastically over the years, given the hype around it. More and more developers are looking for tools that can help them get started quickly with Kubernetes. It also causes some confusion among developers: which tool they should use in order to spend less time configuring their local setup, or writing scripts to automate their inner dev loop workflow? Moreover, developers require better tools while working with Kubernetes because the focus should be on the task at hand, that is, coding, rather than agonizing about how and where they will deploy an application. Ideally, you would prefer a tool that provides extensibility to support various use cases.

This book will show you how to solve inner development loop intricacies in cloud-native applications by automating build, push, and deploy boilerplate using Skaffold.

Who this book is for

This book is for cloud-native application developers, software engineers working with Kubernetes, and DevOps professionals looking for a solution to simplify their inner development loop and improve their CI/CD pipeline for cloud-native applications. Beginner-level knowledge of Java, Docker, Kubernetes, and the containers ecosystem is required before taking on this book.

What this book covers

Chapter 1, *Code, Build, Test, and Repeat - The Application Development Inner Loop*, defines the inner loop of application development and its importance. It also compares the inner with the outer development loop, and covers the typical development workflows for a traditional monolith application and a container-native microservices application.

Chapter 2, *Developing Cloud-Native Applications with Kubernetes – A Developer's Nightmare*, explains the problems that developers face while developing cloud-native applications with Kubernetes.

Chapter 3, *Skaffold – Easy-Peasy Cloud-Native Kubernetes Application Development*, provides a high-level overview of Skaffold. We will also demonstrate Skaffold's basic features by building and deploying a Spring Boot application.

Chapter 4, Understanding Skaffold's Features and Architecture, explores Skaffold's features and internals by looking at its architecture, workflow, and configuration file, `skaffold.yaml`.

Chapter 5, Installing Skaffold and Demystifying Its Pipeline Stages, explains Skaffold installation and common CLI commands used in its different pipeline stages.

Chapter 6, Working with Skaffold Container Image Builders and Deployers, explains various tools used for building (Docker, Jib, kaniko, Buildpacks) and deploying (Helm, kubectl, kustomize) container images to Kubernetes with Skaffold.

Chapter 7, Building and Deploying a Spring Boot Application with the Cloud Code Plugin, introduces you to the Cloud Code plugin developed by Google. It explains how to build and deploy a Spring Boot application to a Kubernetes cluster using the Cloud Code plugin with an IDE such as IntelliJ.

Chapter 8, Deploying a Spring Boot Application to Google Kubernetes Engine Using Skaffold, explains how you can deploy a Spring Boot application to Google Kubernetes Engine, a managed Kubernetes service provided by Google Cloud Platform with Skaffold.

Chapter 9, Creating a Production-Ready CI/CD Pipeline with Skaffold, explains how you can create a production-ready continuous integration and deployment pipeline of a Spring Boot application using Skaffold and GitHub actions.

Chapter 10, Exploring Skaffold Alternatives, Best Practices, and Pitfalls, looks at Skaffold alternative tools such as Telepresence, and also covers Skaffold best practices and traps.

To get the most out of this book

Software/Hardware covered in the book	OS Requirements
OpenJDK 16	Windows, macOS, and Linux
Spring Boot 2.5	
Docker Desktop for macOS and Windows	
IntelliJ or Eclipse IDE	
Skaffold CLI v1.29.0 or above	

If you are using the digital version of this book, we advise you to type the code yourself or access the code via the GitHub repository (link available in the next section). Doing so will help you avoid any potential errors related to the copying and pasting of code.

Download the example code files

You can download the example code files for this book from GitHub at `https://github.com/PacktPublishing/Effortless-Cloud-Native-App-Development-Using-Skaffold`. In case there's an update to the code, it will be updated on the existing GitHub repository.

We also have other code bundles from our rich catalog of books and videos available at `https://github.com/PacktPublishing/`. Check them out!

Download the color images

We also provide a PDF file that has color images of the screenshots/diagrams used in this book. You can download it here:

`https://static.packt-cdn.com/downloads/9781801077118_ColorImages.pdf`

Conventions used

There are a number of text conventions used throughout this book.

`Code in text`: Indicates code words in text, database table names, folder names, filenames, file extensions, pathnames, dummy URLs, user input, and Twitter handles. Here is an example: "Internally, Skaffold creates a `tar` file with changed files that match the sync rules we define in the `skaffold.yaml` file."

A block of code is set as follows:

```
profiles:
  - name: userDefinedPortForward
    portForward:
      - localPort: 9090
        port: 8080
        resourceName: reactive-web-app
        resourceType: deployment
```

Any command-line input or output is written as follows:

```
curl -Lo skaffold https://storage.googleapis.com/skaffold/
releases/latest/skaffold-linux-amd64 && \sudo install skaffold
/usr/local/bin/
```

Bold: Indicates a new term, an important word, or words that you see onscreen. For example, words in menus or dialog boxes appear in the text like this. Here is an example: "Now that we have a working project, click the **Run/Debug Configurations** dropdown and select **Edit Configurations**."

> Tips or important notes
> Appear like this.

Get in touch

Feedback from our readers is always welcome.

General feedback: If you have questions about any aspect of this book, mention the book title in the subject of your message and email us at customercare@packtpub.com.

Errata: Although we have taken every care to ensure the accuracy of our content, mistakes do happen. If you have found a mistake in this book, we would be grateful if you would report this to us. Please visit www.packtpub.com/support/errata, selecting your book, clicking on the Errata Submission Form link, and entering the details.

Piracy: If you come across any illegal copies of our works in any form on the Internet, we would be grateful if you would provide us with the location address or website name. Please contact us at copyright@packt.com with a link to the material.

If you are interested in becoming an author: If there is a topic that you have expertise in and you are interested in either writing or contributing to a book, please visit authors.packtpub.com.

Share Your Thoughts

Once you've read *Effortless Cloud-Native App Development Using Skaffold*, we'd love to hear your thoughts! Scan the QR code below to go straight to the Amazon review page for this book and share your feedback.

https://packt.link/r/1801077118

Your review is important to us and the tech community and will help us make sure we're delivering excellent quality content.

Section 1: The Kubernetes Nightmare – Skaffold to the Rescue

In this section, we will describe the pain and suffering of developing an application with Kubernetes. There are several manual touchpoints in developing a Kubernetes application locally, and it decreases the productivity of a developer. The focus should be on writing code and adding more features to the product rather than worrying about replicating infrastructure on your workstation to debug an issue or test a feature. Engineers at Google described this as an infinite loop of pain and suffering. We will introduce you to Skaffold and how it can help you automate the build, push, and deploy workflow for applications running on Kubernetes.

In this section, we have the following chapters:

- *Chapter 1, Code, Build, Test, and Repeat – The Application Development Inner Loop*
- *Chapter 2, Developing Cloud-Native Applications with Kubernetes – A Developer's Nightmare*
- *Chapter 3, Skaffold – Easy-Peasy Cloud-Native Kubernetes Application Development*

1
Code, Build, Test, and Repeat – The Application Development Inner Loop

Building and deploying cloud-native applications can be cumbersome for local and remote development if you are not using the appropriate tools. Developers go through a lot of pain to automate the build, push, and deploy steps. In this book, we will introduce you to **Skaffold**, which helps automate these development workflow steps. You will learn how to use the Skaffold CLI to accelerate the inner development loop and how to create effective **continuous integration/continuous deployment (CI/CD)** pipelines and perform build and deployment to manage Kubernetes instances such as **Google Kubernetes Engine (GKE)**, **Microsoft's Azure Kubernetes Service (AKS)**, and Amazon's **Elastic Kubernetes Service (EKS)**.

This chapter will define the inner loop for application development and its importance, comparing the inner with the outer development loops, and cover the typical development workflows for a traditional monolith application and a container-native microservices application. We will have an in-depth discussion about the differences between these two approaches.

In this chapter, we're going to cover the following main topics:

- Understanding what the application development inner loop is
- Inner versus outer development loops
- Exploring the traditional application development inner loop
- Checking out the container-native application development inner loop

By the end of this chapter, you will understand the traditional and container-native application inner development loops.

Technical requirements

To follow along with the examples in this chapter, you need the following:

- Eclipse (`https://www.eclipse.org/downloads/`) or IntelliJ IDEA (`https://www.jetbrains.com/idea/download/`)
- Git (`https://git-scm.com/downloads`)
- Spring Boot 2.5 (`https://start.spring.io`)
- minikube (`https://minikube.sigs.k8s.io/docs/`) or Docker Desktop for macOS and Windows (`https://www.docker.com/products/docker-desktop`)
- OpenJDK 16 (`https://jdk.java.net/16/`)

You can download the code examples for this chapter from the GitHub repository at `https://github.com/PacktPublishing/Effortless-Cloud-Native-App-Development-Using-Skaffold/tree/main/Chapter01`

Understanding the application development inner loop

The **application development inner loop** is an iterative process in which a developer changes the code, starts a build, runs the application, and then tests it. If something goes wrong, then we repeat the entire cycle.

So basically, it is the phase before a developer shares the changes done locally with others. Irrespective of your technology stack, the tools used, and personal preferences, the inner loop process may vary, but ideally, it could be summarized into the following three steps:

1. Code
2. Build
3. Test

Here is a quick visual representation of the inner development loop:

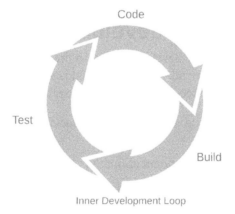

Figure 1.1 – Inner loop

If you think about it, coding is the only step that adds value, and the rest of the steps are like a validation of your work, that is, confirming whether your code is compiling and tests are passing or not. Since developers spend most of their time on the inner loop, they don't like spending too much time on any of the steps. It should be swift. Moreover, as developers, we thrive on fast feedback.

All the steps that we have defined until now are happening locally on a developer's machine before committing the code to a source code repository. Once a developer commits and pushes changes to the source code repository, it typically starts their CI/CD pipeline, called the **outer development loop** (pull request, CI, deployment, and so on). Whether you are developing traditional monolith or container-native microservices applications, you should not neglect the importance of your inner development loop. Here is why you should care about your inner development loop:

- If your inner development loop is slow and lacks automation, then the developer's productivity will plunge.

- It would be best if you always aimed to optimize it because a slow inner loop will affect other dependent teams, and it will take much longer to deliver a new feature to your users.

Now that we've had a quick overview of the application development inner loop, let's compare the inner and outer development loops.

Inner versus outer development loops

As discussed earlier, as long as the developer works in their local environment to test things, they are in the inner loop. In general, a developer spends most of their time in the inner loop because it's fast and gives instant feedback. It usually involves the following steps:

1. A developer starts working on a new feature request. Some code changes are done at this point.
2. Once the developer feels confident about the changes, a build is started.
3. If the build is successful, then the developer runs the unit tests.
4. If the test passes, then the developer starts an instance of the application locally.
5. They will switch to the browser to verify the changes.
6. The developer will then trace logs or attach a debugger.
7. If something breaks, then the developer will repeat the preceding steps.

But as soon as a developer commits and pushes the code to a source code repository, it triggers the outer development loop. The outer development loop is closely related to the CI/CD process. It involves steps such as the following:

1. CI checking out the source code
2. Building the project
3. Running functional and integration test suites

4. Creating runtime artifacts (JAR, WAR, and so on)

5. Deploying to the target environment

6. Testing and repeating

All the preceding steps are typically automated and require minimal to no involvement on the part of a developer. When the CI/CD pipeline breaks because of a test failure or compilation issue, the developer should get notified and then start working again on the inner development loop to fix this issue. Here is a visualization of the inner loop versus the outer loop:

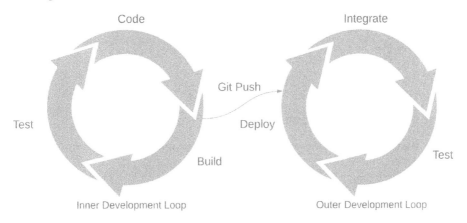

Figure 1.2 – Inner loop versus outer loop

It's very tempting to use CI/CD as a replacement for your inner development loop. Let's discuss whether this is a good approach or not.

Why not use CI/CD?

Contrary to what we just discussed about the inner loop, some developers may say that they don't care about their inner development loop because they have a CI/CD process for it, which should suffice. They are not entirely wrong as these pipelines are purpose-built to make the process of modern application development repeatable and straightforward. Still, your CI/CD process only solves a unique set of problems.

Using CI/CD as a replacement for your inner development loop will make the entire process even slower. Imagine having to wait for the whole CI/CD system to run your build and test suite, and then deploy only to find out that you made a small mistake; it would be quite aggravating. Now, you would have to wait and repeat the entire process just because of some silly mistake. It would be much easier if we can avoid unnecessary iterations. For your inner development loop, you must iterate quickly and preview changes as if they are happening on a live cluster.

We have covered enough basics about the application development inner loop, and now we will cover the traditional application development inner loop for Java developers.

Exploring the traditional application development inner loop

Before containers were cool, we were spoilt by the choices we have for the inner development loop. Your IDE can run builds for you in the background, and then you can deploy your application and test your changes locally. A typical traditional application development inner loop involves steps such as the following:

1. A developer making code changes in an IDE

2. Building and packaging the application

3. Deploying and then running locally on a server

4. Finally, testing the changes and repeating the step

    Here is a visualization of the traditional application development inner loop:

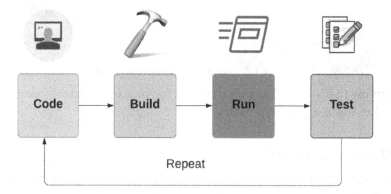

Figure 1.3 – Traditional application development inner loop

For Java developers, there are many options available to automate this process. Some of the most popular options are as follows:

- Spring Boot Developer Tools

- JRebel

Let's discuss these options briefly.

Spring Boot Developer Tools

Spring Boot first introduced developer tools in version 1.3. Spring Boot Developer Tools provide fast feedback loops and automatic restart of the application for any code changes done. It provides the following functionalities:

- It provides a **hot reloading** feature. As soon as any file changes are done on `classpath`, it will automatically reboot the application. The automatic restart may differ based on your IDE. Please check the official documentation (`https://docs.spring.io/spring-boot/docs/1.5.16.RELEASE/reference/html/using-boot-devtools.html#using-boot-devtools-restart`) for more details on this.

- It provides integration with the **LiveReload** plugin (`http://livereload.com`) so that it can refresh the browser automatically whenever a resource is changed. Internally, Spring Boot will start an embedded LiveReload server, which will trigger a browser refresh whenever a resource is changed. The plugin is available for most popular browsers, such as Chrome, Firefox, and Safari.

- It not only supports the local development process, but you can opt-in for updating and restarting your application running remotely on a server or cloud. You can enable remote debugging as well if you like. However, there is a security risk involved in using this feature in production.

The following is a short snippet of how to add relevant dependencies to your Maven and Gradle projects to add support for Spring Boot Developer Tools. Maven/Gradle should have an introduction section first:

Maven pom.xml

```
<dependencies>
  <dependency>
    <groupId>org.springframework.boot</groupId>
    <artifactId>spring-boot-devtools</artifactId>
  </dependency>
</dependencies>
```

Gradle build.gradle

Here is the code for Gradle:

```
dependencies {
compileOnly("org.springframework.boot:spring-boot-devtools")
}
```

But this is not how we will add dependencies to test the auto-reload feature of developer tools. We will use the **Spring Initializr** website (`https://start.spring.io/`) to generate the project stub based on the options you choose. Here are the steps we'll follow:

1. You can go ahead with default options or make your own choices. You can select the build tools (Maven or Gradle), language (Java, Kotlin, or Groovy), and Spring Boot version of your choice.

2. After that, you can add necessary dependencies by clicking on the **ADD DEPENDENCIES...** button and selecting the dependencies required for your application.

3. I have chosen the default options and added `spring-boot-starter-web`, `spring-boot-dev-tools`, and Thymeleaf as dependencies for my demo Hello World Spring Boot application.

4. Now, go ahead and click on the **GENERATE** button to download the generated source code on your computer. Here is the screen you should see:

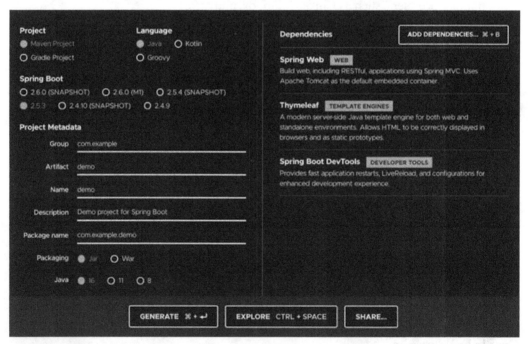

Figure 1.4 – Spring Initializr home page

5. After the download, you can import the project to your IDE.

The next logical step is to build a simple Hello World Spring Boot web application. Let's begin.

Anatomy of the Spring Boot web application

The best way to understand the working parts of the Spring Boot application is by taking a look at an example. In this example, we will create a simple **Spring Web MVC** application that will accept HTTP GET requests at `http://localhost:8080/hello`. We will get an HTML web page with "Hello, John!" in the HTML body in response. We will allow the user to customize the default response by entering the query string in the `http://localhost:8080/hello?name=Jack` URL so that we can change the default message. Let's begin:

1. First, let's create a `HelloController` bean using the `@Controller` annotation for handling incoming HTTP requests. The `@GetMapping` annotation binds the HTTP GET request to the `hello()` method:

```
@Controller
public class HelloController {
    @GetMapping("/hello")
    public String hello(@RequestParam(defaultValue =
      "John", name = "name", required = false) String name,
      Model model) {
        model.addAttribute("name", name);
        return "index";
    }
}
```

This controller returns the name of the view, which is `index` in our case. The view technology we have used here is Thymeleaf, which is responsible for server-side rendering of the HTML content.

2. In the source code template, `index.html` is available under the templates folder in `src/main/resources/`. Here are the contents of the file:

```
<!DOCTYPE HTML>
<html xmlns:th="http://www.thymeleaf.org">
<head>
        <meta charset="UTF-8"/>
        <title>Welcome</title>
</head>
<body>
<p th:text="''Hello, ' + ${name} + '!'" />
</body>
</html>
```

3. Spring Boot provides an opinionated setup for your application, which includes a `main` class as well:

```
@SpringBootApplication
public class Application {
    public static void main(String[] args) {
        SpringApplication.run(Application.class, args);
    }
}
```

4. We will run our application using `mvn spring-boot:run` maven goal, which is provided by `spring-boot-maven-plugin`:

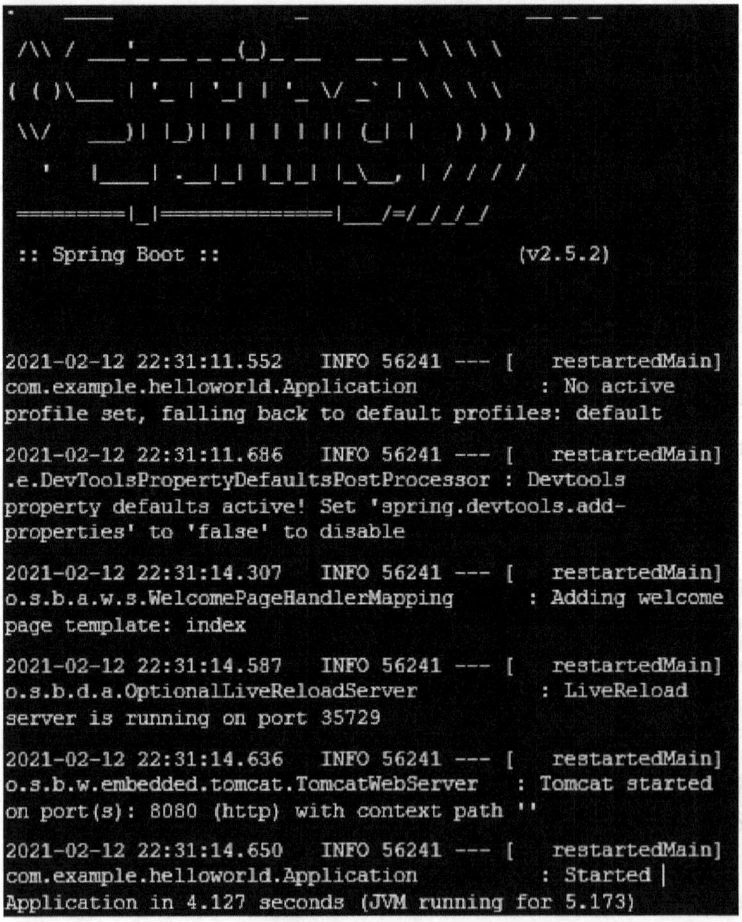

Figure 1.5 – Spring Boot application startup logs

> **Note**
>
> To reduce the verbosity of the logs, we have trimmed them down to show only the parts that are relevant to our discussion.

If you observe the logs carefully, we have developer tools support enabled, an embedded Tomcat server listening at port `8080`, and an embedded LiveReload server running on port `35279`. So far, this looks good. Once the application is started, you can access the `http://localhost:8080/hello` URL.

Hello, John!

Figure 1.6 – REST endpoint response

5. Now we will do a small code change in the Java file and save it and you can see from the logs that the embedded Tomcat server was restarted. In the logs, you can also see that the thread that has spawned the application is not a main thread instead of a `restartedMain` thread:

```
2021-02-12 16:28:54.500    INFO 53622 --- [nio-8080-
exec-1] o.a.c.c.C.[Tomcat].[localhost].[/]          :
Initializing Spring DispatcherServlet 'dispatcherServlet'
2021-02-12 16:28:54.500    INFO 53622 --- [nio-8080-
exec-1] o.s.web.servlet.DispatcherServlet           :
Initializing Servlet 'dispatcherServlet'
2021-02-12 16:28:54.501    INFO 53622 --- [nio-8080-
exec-1] o.s.web.servlet.DispatcherServlet           :
Completed initialization in 1 ms
2021-02-12 16:29:48.762    INFO 53622 --- [
Thread-5] o.s.s.concurrent.ThreadPoolTaskExecutor   :
Shutting down ExecutorService 'applicationTaskExecutor'
2021-02-12 16:29:49.291    INFO 53622 --- [
restartedMain] c.e.helloworld.HelloWorldApplication
: Started HelloWorldApplication in 0.483 seconds (JVM
running for 66.027)
2021-02-12 16:29:49.298    INFO 53622 --- [
restartedMain] .ConditionEvaluationDeltaLoggingListener :
Condition evaluation unchanged
```

```
2021-02-12 16:29:49.318    INFO 53622 --- [nio-8080-
exec-1] o.a.c.c.C.[Tomcat].[localhost].[/]          :
Initializing Spring DispatcherServlet 'dispatcherServlet'
```
```
2021-02-12 16:29:49.319    INFO 53622 --- [nio-8080-
exec-1] o.s.web.servlet.DispatcherServlet            :
Initializing Servlet 'dispatcherServlet'
```
```
2021-02-12 16:29:49.320    INFO 53622 --- [nio-8080-
exec-1] o.s.web.servlet.DispatcherServlet            :
Completed initialization in 1 ms
```

This completes the demo of the auto-restart feature of the Spring Boot Developer Tools. We have not covered the LiveReload feature for brevity as it would be difficult to explain here because it all happens in real time.

JRebel

JRebel (https://www.jrebel.com/products/jrebel) is another option for Java developers for accelerating their inner loop development process. It is a JVM plugin, and it helps in reducing time for local development steps such as building and deploying. It is a paid tool developed by a company named *Perforce*. However, there is a free trial for 10 days if you would like to play with it. It provides the following functionalities:

- It allows developers to skip rebuild and redeploys and see live updates of their changes by just refreshing the browser.

- It will enable developers to be more productive while maintaining the state of their application.

- It provides an instant feedback loop, which allows you to test and fix your issues early in your development.

- It has good integration with popular frameworks, application servers, build tools, and IDEs.

There are many different ways to enable support for JRebel to your development process. We will consider the possibility of using it with an IDE such as Eclipse or IntelliJ. For both IDEs, you can install the plugin, and that's it. As I said earlier, this is a paid option, and you can only use it for free for 10 days.

For IntelliJ IDEA, you can install the plugin from the marketplace.

Figure 1.7 – IntelliJ IDEA installing JRebel

For the Eclipse IDE, you can install the plugin from Eclipse Marketplace.

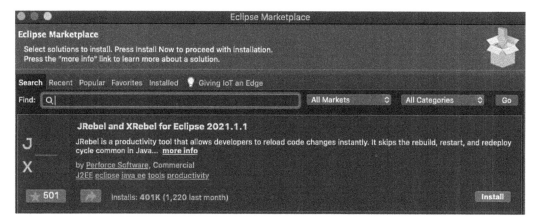

Figure 1.8 – Eclipse IDE installing JRebel

Since JRebel is a paid option, we will not be exploring it in this book, but you are free to test it yourself.

We have covered the traditional application development inner loop life cycle and tools such as Spring Boot Developer Tools and JRebel, which allow rapid application development. Let's now go through the container-native application development inner loop life cycle.

Checking out the container-native application development inner loop

Kubernetes and containers have introduced a new set of challenges and complexities to the inner development loop. Now there are an additional set of steps added to the inner loop while developing applications, which is time-consuming. A developer would prefer to spend time solving business problems rather than waiting for the build process to complete.

It involves steps such as the following:

1. A developer making code changes in an IDE
2. Building and packaging the application
3. Creating a container image
4. Pushing the image to a container registry
5. Kubernetes pulling the image from the registry
6. Kubernetes creating and deploying the pod
7. Finally, testing and repeating

Engineers at Google call this an *infinite loop of pain and suffering*. Here is a visualization of the container-native application development inner loop:

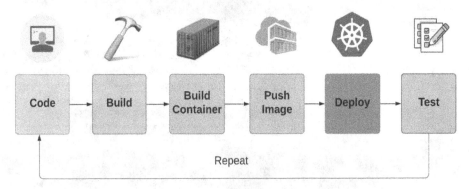

Figure 1.9 – Container-native application development inner loop

As you can see, we now have three more steps added to the inner development loop, that is, creating a container image of your application, pushing it to a container registry, and finally, pulling the image while deploying to a container orchestration tool such as Kubernetes.

The container image could be a Docker or OCI format image, depending on the tool you use to build your images. You have options such as Docker Hub, AWS Container Registry, Google Container Registry, or Azure Container Registry for the container registry. Then, finally, in deployment, for your container orchestration, you have tools such as Kubernetes, which will first pull the image from the container registry and deploy your application.

There are many manual steps involved here. It also depends on what tools you have used for the local development workflow. For instance, you will use commands such as the following:

```
docker build
docker tag
docker push
kubectl apply
```

The following are the detailed steps that a developer has to go through while developing container-native applications:

1. Defining how to configure the OS for your container with a Dockerfile

2. Defining the packaging of your application into a container image by adding instructions to the Dockerfile

3. Creating a container image with Docker commands such as `docker build` and `docker tag`

4. Uploading the container image to a container registry with a command such as `docker push`

5. Writing one or more Kubernetes resource files in YAML

6. Deploying your application to the cluster with commands such as `kubectl apply -f myapp.yaml`

7. Deploying services to the cluster with commands such as `kubectl apply -f mysvc.yaml`

8. Writing the config so that apps can work together with commands such as `kubectl create configmap`

9. Configuring apps to work together correctly with commands such as `kubectl apply -f myappconfigmap.yaml`

Wooh!!! That's a lot of steps and a time-consuming process. You can use scripting or `docker compose` to automate it to some extent, but soon you will realize that it cannot be fully automated without a tool such as Skaffold, which can abstract away many things related to building and deployment.

In *Chapter 3, Skaffold – Easy-Peasy Cloud-Native Kubernetes Application Development*, we will cover Skaffold, which simplifies the process we have covered here with a single command. My only intention here was to give you an idea of the steps involved. We will cover these steps with some hands-on examples in the next chapter.

Summary

In this chapter, we have covered many topics, such as what a typical inner development loop is and its importance. We have also discussed how both the inner and outer development loops are different, and then we explored whether the CI/CD process can act as a replacement for the inner development loop.

We then discussed the steps involved in the traditional application development inner loop and we covered tools such as Spring Developer Tools and JRebel, which make the application development a lot easier. To explain this further, we created a simple Spring Boot web MVC application. Finally, in the last section, we covered the container-native application development inner loop. We also covered the steps involved in container-native application development.

In this chapter, the focus was on introducing you to concepts such as inner and outer development. You can use Spring Boot Developer Tools and JRebel to accelerate/automate your traditional application development life cycle.

In the next chapter, we will cover the problems a developer faces while developing an application with Kubernetes.

Further reading

- Learn more about Spring Boot Developer Tools at `https://docs.spring.io/spring-boot/docs/1.5.16.RELEASE/reference/html/using-boot-devtools.html`.

- More information on JRebel is available at `https://www.jrebel.com/`.

- Learn more about Docker from *Docker for Developers,* published by Packt (`https://www.packtpub.com/product/docker-for-developers/9781789536058`).

- Learn more about Kubernetes from *Mastering Kubernetes,* published by Packt (`https://www.packtpub.com/product/mastering-kubernetes/9781786461001`)

2
Developing Cloud-Native Applications with Kubernetes – A Developer's Nightmare

In the previous chapter, we covered the hardships that a developer faces while developing a container-native application. We also covered the new steps that were introduced to the development life cycle. We may have simplified things to explain the concepts, but we will look into each step in detail in this chapter.

This chapter will cover the problems that developers face while developing cloud-native applications with Kubernetes. We will cover how and why the overall development experience with Kubernetes is painful. We will also see why developers are not Kubernetes experts and that they are looking for simplified workflows while developing an application with Kubernetes.

In this chapter, we're going to cover the following main topics:

- Poor developer experience
- Developers want simplified workflows with Kubernetes
- Developers are not Kubernetes experts

By the end of this chapter, you will understand the common challenges developers face while developing cloud-native applications with Kubernetes. Later, in the next chapter, we will learn how to overcome these challenges by using Skaffold for your development workflow.

Technical requirements

To follow along with the examples in this chapter, you need the following:

- Eclipse (`https://www.eclipse.org/downloads/`) or IntelliJ IDE (`https://www.jetbrains.com/idea/download/`)
- Git (`https://git-scm.com/downloads`)
- Spring Boot 2.5 (`https://start.spring.io`)
- minikube (`https://minikube.sigs.k8s.io/docs/`) or Docker Desktop for macOS and Windows (`https://www.docker.com/products/dockerdesktop`)
- OpenJDK 16 (`https://jdk.java.net/16/`)

You can download the code examples for this chapter from the GitHub repository at `https://github.com/PacktPublishing/Effortless-Cloud-Native-App-Development-Using-Skaffold/tree/main/Chapter02`.

Poor developer experience

Modern developers are looking for tools and technologies that give them an edge to deliver the software that meets the customers' expectations and keeps their organization competitive in today's fast-moving world. Enter Kubernetes! **Kubernetes** was open sourced in 2014, and since its inception, it has become the container orchestration platform of choice for numerous enterprises around the world. Kubernetes has dramatically simplified the jobs of operations folks, but we definitely cannot say the same thing about developers who are building and deploying applications to Kubernetes.

We have covered this in detail in this chapter. According to a recent study, around 59 percent of enterprise organizations are running their production workload with Kubernetes. This is excellent for a technology that is just 5 years old. The main reason enterprises are adopting Kubernetes is to increase agility, expedite software delivery, and support digital transformation.

Before going through the pain points of working with Kubernetes, let's take a real-world example to understand how Kubernetes can help organizations in their digital transformation journey. Let's take an example of an e-commerce website. On most days, the website does its job. The website takes advantage of microservices architecture and has multiple services that work in tandem to deliver a better user experience. However, due to an upcoming holiday, the IT team expects a surge in the usage of their website, and the team is worried that it may lead to an outage, as the underlying microservices may not be able to handle the load. But with Kubernetes, it is easy to scale out without much hassle. For example, you can use Kubernetes' autoscaling capabilities with its **horizontal Pod auto-scaler** (HPA). HPA automatically scales the number of Pods based on observed CPU utilization.

Furthermore, containers and Kubernetes certainly have changed the way we package, deploy, and run cloud-native applications at scale. After containerization, you can run your application anywhere, that is, on VMs, on physical machines, or the cloud. And with container orchestration tools such as Kubernetes, you can scale, deploy, and manage cloud-native applications more efficiently. It has reduced the downtime in production and made the job of the operations team much more comfortable. However, developer experience and practices have not evolved much since the inception of Kubernetes as compared to traditional applications. Let's understand cloud-native application development flow with an example.

Understanding the cloud-native application development workflow

We will use the same *Hello-World Spring Boot Web MVC* application we created in *Chapter 1, Code, Build, Test, and Repeat – The Application Development Inner Loop*; however, we will containerize it and deploy it to Kubernetes this time. The idea is to go through the hardships that a developer goes through while developing cloud-native Spring Boot applications. Here are the steps we'll be following:

1. We will be using **Docker Desktop** for macOS and Windows as it comes with Kubernetes support, and we will not have to download **minikube** separately for this example. However, if macOS is not something you work with, then you can install minikube (`https://v1-18.docs.kubernetes.io/docs/tasks/tools/install-minikube/#installing-minikube`) for other OSes as well. Follow the steps to enable Kubernetes support with Docker Desktop for macOS and Windows.

2. Navigate to **Preferences** in the Docker menu bar. Then, on the **Kubernetes** tab, click on the **Enable Kubernetes** checkbox to start a single-node functional Kubernetes cluster. It will take some time to start the cluster. It is not mandatory, but you can also enable Kubernetes to be the default orchestrator for the `docker stack` command.

Figure 2.1 – Enabling Kubernetes

3. After it is enabled, you will see the following screen on your Docker Desktop menu bar. This confirms that the Kubernetes cluster is up and running:

Figure 2.2 – Verifying the setup

4. Next, make sure that the Kubernetes context is set to `docker-desktop` if you have multiple clusters or environments running locally:

Figure 2.3 – Context set to docker-desktop

5. By the way, Docker Desktop comes with **kubectl** support; you don't have to download it separately. kubectl is a command-line tool for Kubernetes, and you can use it to run commands against your cluster. On macOS, it's generally available at path, `/usr/local/bin/kubectl`. For Windows, it is available at `C:\>Program Files\Docker\Docker\Resources\bin\kubectl.exe`. You may want to add it to your `PATH` variable. Let's verify the setup with the following command:

```
kubectl get nodes
NAME             STATUS    ROLES     AGE    VERSION
docker-desktop   Ready     master    59d    v1.19.3
```

6. The following is the Dockerfile we have used for this example:

```
FROM openjdk:16
COPY target/*.jar app.jar
ENTRYPOINT ["java","-jar","/app.jar"]
```

We have a very basic Dockerfile here. Let me just explain the instructions in brief:

a. The `FROM` instruction specifies the base image for our Spring Boot application, which is OpenJDK 16.

b. `COPY` is used to move files or directories from the host system to the filesystem inside the container. Here, we have copied the `.jar` files from the target directory to the root path inside the container.

c. `ENTRYPOINT` works as a runtime executable for the container, which will start our application.

7. Now that we have the Dockerfile, next we need to create an executable .jar file. We will use the mvn clean install command to create an executable .jar file for our application. Let's run the docker build command to create a container image. Here, we have set the name of our image as helloworld. The output of the docker build command will be the following:

```
docker build -t hiashish/helloworld:latest .
[+] Building 4.9s (8/8) FINISHED
 => [internal] load build definition from Dockerfile
0.1s
 => => transferring dockerfile: 36B
0.0s
 => [internal] load .dockerignore
0.0s
 => => transferring context: 2B
0.0s
 => [internal] load metadata for docker.io/library/
openjdk:16
4.3s
 => [auth] library/openjdk:pull token for registry-1.
docker.io
0.0s
 => [internal] load build context
0.1s
 => => transferring context: 86B
0.0s
 => [1/2] FROM docker.io/library/openjdk:11@sha256:
3805f5303af58ebfee1d2f5cd5a897e97409e48398144afc223
3f7b778337017
0.0s
 => CACHED [2/2] COPY target/*.jar app.jar
0.0s
 => exporting to image
0.0s
 => => exporting layers
0.0s
 => => writing image sha256:65b544ec877ec10a4dce9883b3
766fe0d6682fb8f67f0952a41200b49c8b0c50
```

```
0.0s
 => => naming to docker.io/hiashish/helloworld:latest
```

8. We have created an image for the application. Now we are ready to push the image to the DockerHub container registry with the docker push command, as follows:

```
docker push hiashish/helloworld
Using default tag: latest
The push refers to repository [docker.io/hiashish/
helloworld]
7f517448b554: Pushed
ebab439b6c1b: Pushed
c44cd007351c: Pushed
02f0a7f763a3: Pushed
da654bc8bc80: Pushed
4ef81dc52d99: Pushed
909e93c71745: Pushed
7f03bfe4d6dc: Pushed
latest: digest: sha256:16d3d9db1ecdbf21c69bc838d4a
a7860ddd5e212a289b726ac043df708801473 size: 2006
```

9. The last part of this exercise is to create Kubernetes resources (Deployments and Services) to get our application up and running on Kubernetes. The declarative YAML file for both services and deployment is inside the K8s directory of the source code. Let's create the Deployment resource first, which is responsible for creating and running a set of Pods dynamically:

```
apiVersion: apps/v1
kind: Deployment
metadata:
  labels:
    app: helloworld
  name: helloworld
spec:
  replicas: 1
  selector:
    matchLabels:
      app: helloworld
  template:
```

```
metadata:
  labels:
    app: helloworld
spec:
  containers:
    - image: docker.io/hiashish/helloworld
      name: helloworld
```

Let me clarify few things regarding the YAML file we have used to create the Kubernetes Deployment object:

a. `metadata.name` specifies the name of the Deployment object to be created.

b. The `spec.replicas` field indicates that the Kubernetes Deployment object will create a single replica.

c. The `template.spec` field indicates that the Pod will run a single container named `helloworld` that runs the DockerHub image of our application.

Here is the `kubectl` command for creating the Deployment object:

```
kubectl create -f mydeployment.yaml
deployment.apps/helloworld created
```

10. Services provide a single DNS name for a set of Pods and handle load balancing among them. Let's create the Service resource so that the application can be accessed from outside the cluster:

```
apiVersion: v1
kind: Service
metadata:
  labels:
    app: helloworld
  name: helloworld
spec:
  ports:
    - port: 8080
      protocol: TCP
      targetPort: 8080
  selector:
    app: helloworld
  type: NodePort
```

Let's talk about the YAML file we have used to create the Kubernetes Service object:

a. `metadata.name` specifies the name of the Service object to be created.

b. `spec.selectors` allows Kubernetes to group Pods with the name `helloworld` and forward the request to them.

c. `type: Nodeport` creates a static IP for each node so that we can access the Service from outside.

d. `targetPort` is the container port.

e. `port` is the port exposed internally in the cluster.

The following is the `kubectl` command for creating the Service object:

```
kubectl create -f myservice.yaml
service/helloworld created
```

11. Let's now verify whether we have a Pod running:

```
NAME                                READY    STATUS     RESTARTS    AGE
pod/helloworld-7fb77f6977-bbj7v     1/1      Running    0           118s

NAME                     TYPE         CLUSTER-IP        EXTERNAL-IP     PORT(S)            AGE
service/helloworld       NodePort     10.111.255.135    <none>          8080:31904/TCP     103s
service/kubernetes       ClusterIP    10.96.0.1         <none>          443/TCP            44h

NAME                              READY    UP-TO-DATE    AVAILABLE    AGE
deployment.apps/helloworld        1/1      1             1            118s

NAME                                         DESIRED    CURRENT    READY    AGE
replicaset.apps/helloworld-7fb77f6977        1          1          1        118s
```

Figure 2.4 – Pod running

12. As you can see, we now have our application up and running on Kubernetes. Let's verify this:

Hello, John!

Figure 2.5 – REST endpoint response

That's a lot of steps and too many keystrokes even if your change is small and you don't even know whether it will work or not. Now imagine having to do this every time you push a change! This workflow could be even more complicated if you have multiple microservices talking to each other. You have the option to not deploy to Kubernetes for your local development, but rely instead on your CI/CD process. Or maybe you are using something like `docker-compose` or testing in isolation with Docker. Just imagine having multiple microservices that you need to run this way.

To test everything, realistically, you need your development environment to mirror your deployment environment to test your microservices' dependencies. This is the downside of container-native development, as the developer spends less time coding and more time worrying about the configuration, setting up the environment, and waiting for the deployment to be completed. In one of the chapters, later in the book, we will cover how we can build and deploy multiple microservices with Skaffold.

Due to the inherent complexity that comes with Kubernetes, developers are looking for simple workflows. Let's discuss this in the next section.

Developers want simplified workflows with Kubernetes

In the last chapter, we discussed the steps that a developer goes through while developing traditional Spring Boot applications in the inner development loop. We also discussed how easy it is to automate the whole flow with tools such as *spring-dev-tools*. Once a developer is confident about the changes, they can save them, and changes are deployed automatically.

Developers developing cloud-native applications are looking for a similar workflow where they can save their changes. With some magic in the background, the application should be deployed to local or remote clusters of their choice. Moreover, a developer who has previously worked on traditional monolithic applications would expect a similar workflow when they switch to developing cloud-native applications. From a developer's perspective, the expectation is that additional steps for cloud-native application development should be suppressed with a single command or click.

A developer expects a simplified workflow with Kubernetes, as shown in the following diagram:

Figure 2.6 – Ctrl + S workflow with Kubernetes

To address these problems, enterprises need to provide developers with tools that can abstract general Kubernetes complexity. To be specific, developers are looking for a platform or tools that can fulfill the following requirements:

- Developers should be able to connect with Kubernetes without going through the bureaucracy of getting approval from support managers.

- Developers should not be wasting their time and energy configuring the environment.

- Developers should be able to start quickly while working with Kubernetes.

- Developers can quickly deploy the changes to the Kubernetes cluster with a single command.

- Developers should be debugging cloud-native applications during development, such as how they are used to debug traditional applications.

Developers should not be tied to a tool for building and deploying the image. The good news is that many enterprises have realized how painful the developer experience is with Kubernetes and are coming up with their solutions to improve it. Later in this book, we will cover a tool, Skaffold, that simplifies developers' inner development loops while working with cloud-native applications. Skaffold implements the *Ctrl + Save* workflow and automates the build and deploy process. Skaffold also gives the developer the freedom to pick a tool for the build (Docker/Jib/Buildpacks) and deployment (kubectl/Helm/kustomize).

It would be a good skill set to have, but do we really want developers to be Kubernetes experts? Let's discuss this in the next section.

Developers are not Kubernetes experts

Kubernetes was originally developed for operations folks and not for developers. There are many reasons why a developer would not be interested in knowing Kubernetes for their day jobs. One valid reason is that a developer is more interested in solving the business problem and adding features to the products they are developing, and they are not bothered about the target environment, that is, where they will deploy the application. And, let's be honest, Kubernetes is complex, which makes it hard not only for the beginner but also for experienced folks. I saw this joke, probably on Twitter, on how hard it is to understand Kubernetes: *"One time I tried to explain Kubernetes to someone. Then we both didn't understand it."*

It requires a different level skill set than the everyday tasks of a developer. Because of its complexity, it usually takes a very long time for the average developer to master Kubernetes.

More often than not, a developer working in an enterprise environment will be working on the following tasks:

- Being involved in design discussions
- Adding new features for the product
- Writing unit test cases
- Improving code quality
- Working on improving the performance of the application
- Fixing bugs
- Refactoring code

Developers just want to code rather than worry about how and where their applications will be deployed.

The bottom line is that we need to keep telling ourselves that Kubernetes is not an easy tool for developers. Moreover, developers are more interested in creating applications, working with tools that can handle the build, and deploying boilerplate for them.

Summary

This chapter has covered the hardships that a developer has to go through while developing cloud-native applications with Kubernetes. We started the chapter by describing the cloud-native application development workflow for an application deployed to Kubernetes. We covered the additional steps with some coding examples that a developer has to go through while developing cloud-native applications. Then we explained that developers are looking for a simplified workflow for easy development with Kubernetes. Later in the chapter, we showed that developers are not Kubernetes experts, and they should be equipped with tools such as Skaffold to improve their development experience with Kubernetes.

In this chapter, the main goal was to give you a walk-through of developers' problems while developing container-native applications. After reading this, you should be able to relate to these issues, and at the same time, I have given you hints about how Skaffold can help solve these problems.

In the next chapter, we will quickly cover Skaffold with some coding examples to better understand these hints.

Further reading

- Learn more about Docker and Kubernetes at `https://www.packtpub.com/product/kubernetes-and-docker-an-enterprise-guide/9781839213403`.

- More on going cloud-native with Kubernetes can be discovered at `https://www.packtpub.com/product/cloud-native-with-kubernetes/9781838823078`.

3

Skaffold — Easy-Peasy Cloud-Native Kubernetes Application Development

In the previous chapter, we learned that developing applications with Kubernetes is cumbersome and provided some coding examples. This chapter will cover a high-level overview of Skaffold. You will also learn and understand Skaffold basic **command-line interface** (**CLI**) commands and how these ease developers' pain in developing cloud-native microservices with Skaffold. We will demonstrate Skaffold's basic features by building and deploying a Spring Boot application.

In this chapter, we're going to cover the following main topics:

- What is Skaffold?
- Building and deploying a Spring Boot application with Skaffold

By the end of this chapter, you will have a basic understanding of Skaffold and will be able to take advantage of Skaffold to accelerate an inner development loop while developing cloud-native applications.

Technical requirements

To follow along with the examples of this chapter, you will need the following:

- An Eclipse (`https://www.eclipse.org/downloads/`) or IntelliJ IDEA `https://www.jetbrains.com/idea/download/`
- Git
- Skaffold CLI (`https://skaffold.dev/docs/install/`)
- Spring Boot 2.5 (`https://start.spring.io`)
- OpenJDK 16 (`https://jdk.java.net/16/`)
- minikube (`https://minikube.sigs.k8s.io/docs/`) or Docker Desktop for macOS and Windows (`https://www.docker.com/products/dockerdesktop`)

You can download the code examples for this chapter from the GitHub repository at `https://github.com/PacktPublishing/Effortless-Cloud-Native-App-Development-Using-Skaffold/tree/main/Chapter03`.

What is Skaffold?

As most developers, Matt Rickard, a Google engineer, also experienced the same pain while building and deploying Kubernetes applications in the inner loop. Matt decided to take the matter into his own hands and created Skaffold.

Skaffold is a CLI tool that automates the build, push, and deploy steps for cloud-native applications running on local or remote Kubernetes clusters of your choice. Skaffold is not a replacement for Docker or Kubernetes. It works in conjunction with them and handles the build, push, and deploy boilerplate part for you.

Skaffold is an open source tool developed by Google. It was generally available on November 7, 2019, and is released under the Apache 2.0 license. Skaffold is written in the Go programming language. You can visit the Skaffold home page at `https://skaffold.dev/`. Skaffold documentation is available at `https://skaffold.dev/docs/`.

If you are on macOS, then you can use the `homebrew` package manager to install Skaffold with the `brew install skaffold` command. However, we will cover various ways to install Skaffold in *Chapter 5*, *Installing Skaffold and Demystifying Its Pipeline Stages*.

Skaffold is widely popular among the developer community because it provides sensible defaults, is simple to use, and has a pluggable architecture. Here's a recent tweet from the official Skaffold account, just confirming this:

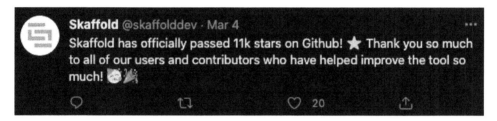

Figure 3.1 – Skaffold Twitter account tweets on passing 11k stars on GitHub

As mentioned in the tweet, the number of stars and forks for the Skaffold GitHub repository speaks for its popularity itself, as we can see here:

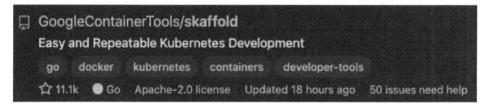

Figure 3.2 – Skaffold GitHub repository

The Skaffold GitHub page is available at `https://github.com/GoogleContainerTools/skaffold`.

Let's try to understand the working of Skaffold by building and deploying a Spring Boot application.

Building and deploying a Spring Boot application with Skaffold

To understand Skaffold commands and concepts better, in this section, we will build and deploy a Spring Boot Java application to a local single-node Kubernetes cluster using Skaffold.

> **Note**
>
> Whenever we talk about *the Kubernetes cluster for local development* in this book, we refer to *the Kubernetes cluster with Docker Desktop*, if not specified otherwise. However, Docker Desktop or minikube is not the only tool available today for running a local Kubernetes cluster. Skaffold also supports Kind https://github.com/kubernetes-sigs/kind and k3d https://github.com/rancher/k3d as target Kubernetes clusters for local development.

Since this will be a sneak-peek of Skaffold, we will not cover everything in detail about Skaffold, as we are going to cover this in the upcoming chapters. I will, however, try to explain the commands used so that you can understand the exact flow. Before we dive into Skaffold, let's first talk about the Spring Boot application we are going to build and deploy with Skaffold.

Creating a Spring Boot application

This Spring Boot application we will be creating will have two **Representational State Transfer (REST)** endpoints exposed. The /states REST endpoint will return all Indian states and their capitals, and the /state?name=statename REST endpoint will return a specific Indian state and its capital. This application uses an in-memory H2 database that inserts rows at the start of the application. Similar to previous chapters, we will use https://start.spring.io to generate stubs for the project. The following screenshot shows the dependencies we are going to use to build this application:

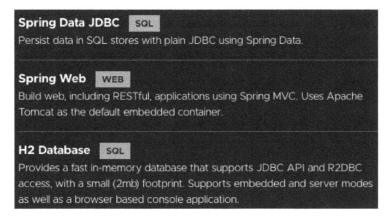

Figure 3.3 – Required dependencies for the Spring Boot application

Add the following dependency to the Maven pom.xml file:

```
<dependency>
    <groupId>org.springframework.boot</groupId>
    <artifactId>spring-boot-starter-data-jdbc</artifactId>
</dependency>
<dependency>
    <groupId>org.springframework.boot</groupId>
    <artifactId>spring-boot-starter-web</artifactId>
</dependency>
<dependency>
    <groupId>com.h2database</groupId>
    <artifactId>h2</artifactId>
    <scope>runtime</scope>
</dependency>
<plugin>
    <groupId>com.google.cloud.tools</groupId>
    <artifactId>jib-maven-plugin</artifactId>
    <version>2.8.0</version>
```

```
<configuration>
  <from>
    <image>openjdk:16-jdk-alpine</image>
  </from>
  <to>
    <image>docker.io/hiashish/skaffold-introduction
      </image>
  </to>
</configuration>
</plugin>
```

Apart from the dependencies we already discussed, I have added the `jib-maven-plugin` plugin to `pom.xml`, which will containerize the Spring Boot application to a container image. Jib takes the input as your source code and outputs a ready-to-run container image of your application. By the way, there's a Gradle equivalent as well. For Gradle, use the following code:

```
plugins {
  id 'com.google.cloud.tools.jib' version '2.8.0'
}
```

> **Tip**
>
> **Jib** can create an image without a Docker daemon. This means that you don't have to install and configure Docker and create or maintain a Dockerfile.
>
> We will cover more about Jib in *Chapter 6, Working with Skaffold Container Image Builders and Deployers*.

Let's begin, then, as follows:

1. This is the layout of the source code directory:

```
.
├── README.md
├── k8s
│   ├── mydeployment.yaml
│   └── myservice.yaml
├── mvnw
├── mvnw.cmd
├── pom.xml
├── skaffold.yaml
└── src
    ├── main
    │   ├── java
    │   │   └── com
    │   │       └── packt
    │   │           └── chapter3
    │   │               └── indianstates
    │   │                   ├── IndianStatesApplication.java
    │   │                   ├── State.java
    │   │                   ├── StateController.java
    │   │                   ├── StateRepository.java
    │   │                   └── StateService.java
    │   └── resources
    │       ├── application.properties
    │       ├── data.sql
    │       └── schema.sql
    └── test
        └── java
            └── com
                └── packt
                    └── chapter3
                        └── indianstates
                            └── IndianStatesApplicationTests.java
```

Figure 3.4 – Project layout

2. The following is the REST controller class annotated with `@RestController`
    annotation for handling incoming **HyperText Transfer Protocol (HTTP)** requests.
    The `@GetMapping` annotations on the `getAllStates()` method binds all
    HTTP GET requests to it when the `/states` REST endpoint is accessed. Similarly,
    the `getSpecificState()` method handles the HTTP GET request for `/state`
    when the state name is passed as a query parameter into the REST **Uniform
    Resource Locator (URL)**. If no parameter is passed, then it takes a default value of
    the `Maharashtra` state:

```
import org.slf4j.Logger;
import org.slf4j.LoggerFactory;
import org.springframework.web.bind.annotation.
GetMapping;
import org.springframework.web.bind.annotation.
RequestParam;
import org.springframework.web.bind.annotation.
RestController;


import java.util.List;


@RestController
public class StateController {
    private final StateService stateService;
    private static final Logger LOGGER =
    LoggerFactory.getLogger(Controller.class);


    public StateController(StateService stateService) {
        this.stateService = stateService;
    }


    @GetMapping("/states")
    private List<State> getAllStates() {
        LOGGER.info("Getting all state");
        return stateService.findAll();
    }


    @GetMapping(value = "/state")
```

```
        private String getSpecificState(@
          RequestParam(required = false, name = "name",
            defaultValue = "Maharashtra") String name) {
            return stateService.findByName(name);
      }
    }
```

3. At the time of writing this book, Java 16 is generally available. I have taken the liberty of also introducing you to some of its cool new features. Let's now talk about records. We have the following data carrier record class:

```
    public record State(String name, String capital) {}
```

The class type is record, and it's a special type that got added as a feature in Java 16. As per the *Java Enhancement Proposal 395* (https://openjdk.java.net/jeps/395), records are a new kind of class in the Java language. They act as transparent carriers for immutable data, with less ceremony than for normal classes. Records can be thought of as nominal tuples. The record class declaration consists of a name, optional type parameters, a header, and a body. Another interesting feature worth mentioning about the record class is that hashcode(), equals(), toString(), and a canonical constructor are implicitly generated for us by the compiler.

4. The following is the StateRepository interface that is implemented by the StateService class:

```
    import java.util.List;
    public interface StateRepository {
        List<State> findAll();
        String findByName(String name);
    }
    import org.springframework.jdbc.core.JdbcTemplate;
    import org.springframework.jdbc.core.RowMapper;
    import org.springframework.stereotype.Service;

    import java.util.List;


    @Service
    public class StateService implements StateRepository{
        private final JdbcTemplate;
```

```java
public StateService(JdbcTemplate jdbcTemplate) {
    this.jdbcTemplate = jdbcTemplate;
}

private final RowMapper<State> rowMapper = (rs,
rowNum) -> new State(rs.getString("name"),
rs.getString("capital"));


@Override
public List<State> findAll() {
    String findAllStates = """
            select * from States
            """;
    return jdbcTemplate.query(findAllStates,
        rowMapper);
}


@Override
public String findByName(String name) {
    String findByName = """
            select capital from States where name
              = ?;
            """;
    return jdbcTemplate.queryForObject(findByName,
        String.class, name);
}
}
```

In the `StateService` class, we are using Spring's `JdbcTemplate` to access the H2 database. The `findAll()` method returns all the states and their capitals. In the same class as the `findAll()` method, I have used the `RowMapper` functional interface. `JdbcTemplate` uses this for mapping rows of a `ResultSet` object and returns a `Row` object for the current row.

I'm sure you may have also observed that I have additionally used the new keyword to initialize the `record` class, which means I can initialize the `record` class like I would a normal class in Java. The `findByName()` method returns a `String`, which is the capital of the state that comes in the `query` parameter request.

I have also used the *Java 15 Text Blocks* (`https://openjdk.java.net/jeps/378`) feature while declaring the **Structured Query Language** (**SQL**) queries, which helps in the readability of SQL queries and **JavaScript Object Notation** (**JSON**) string values.

5. As I explained earlier, we have used the in-memory H2 database to hold the data while the application is running. It gets inserted at the application startup using the following SQL statements:

```
INSERT INTO States VALUES ('Andra Pradesh','Hyderabad');
INSERT INTO States VALUES ('Arunachal
Pradesh','Itangar');
INSERT INTO States VALUES ('Assam','Dispur');
INSERT INTO States VALUES ('Bihar','Patna');
INSERT INTO States VALUES ('Chhattisgarh','Raipur');
INSERT INTO States VALUES ('Goa','Panaji');
INSERT INTO States VALUES ('Gujarat','Gandhinagar');
INSERT INTO States VALUES ('Haryana','Chandigarh');
INSERT INTO States VALUES ('Himachal Pradesh','Shimla');
INSERT INTO States VALUES ('Jharkhand','Ranchi');
INSERT INTO States VALUES ('Karnataka','Bengaluru');
INSERT INTO States VALUES ('Kerala','Thiruvananthapuram');
INSERT INTO States VALUES ('Madhya Pradesh','Bhopal');
INSERT INTO States VALUES ('Maharashtra','Mumbai');
INSERT INTO States VALUES ('Manipur','Imphal');
INSERT INTO States VALUES ('Meghalaya','Shillong');
INSERT INTO States VALUES ('Mizoram','Aizawl');
INSERT INTO States VALUES ('Nagaland','Kohima');
INSERT INTO States VALUES ('Orissa','Bhubaneshwar');
INSERT INTO States VALUES ('Rajasthan','Jaipur');
INSERT INTO States VALUES ('Sikkim','Gangtok');
INSERT INTO States VALUES ('Tamil Nadu','Chennai');
INSERT INTO States VALUES ('Telangana','Hyderabad');
INSERT INTO States VALUES ('Tripura','Agartala');
INSERT INTO States VALUES ('Uttarakhand','Dehradun');
INSERT INTO States VALUES ('Uttar Pradesh','Lucknow');
INSERT INTO States VALUES ('West Bengal','Kolkata');
INSERT INTO States VALUES ('Punjab','Chandigarh');
```

6.   Data is defined using the following schema:

```
DROP TABLE States IF EXISTS;
CREATE TABLE States(name VARCHAR(255), capital
VARCHAR(255));
```

7.   The Kubernetes manifests—that is, deployment and service—are available under the
     k8s directory in the source code, as illustrated in the following code snippet:

mydeployment.yaml

```yaml
apiVersion: apps/v1
kind: Deployment
metadata:
  labels:
    app: skaffold-introduction
  name: skaffold-introduction
spec:
  replicas: 1
  selector:
    matchLabels:
      app: skaffold-introduction
  template:
    metadata:
      labels:
        app: skaffold-introduction
    spec:
      containers:
        - image: docker.io/hiashish/skaffold-introduction
          name: skaffold-introduction
```

myservice.yaml

```yaml
apiVersion: v1
kind: Service
metadata:
  labels:
    app: skaffold-introduction
  name: skaffold-introduction
```

```
  spec:
    ports:
      - port: 8080
        protocol: TCP
        targetPort: 8080
    selector:
      app: skaffold-introduction
    type: LoadBalancer
```

So far, we have covered all the required building blocks for Skaffold. Now, let's talk about the Skaffold configuration.

Understanding the Skaffold configuration

Let's talk about the `skaffold.yaml` Skaffold configuration file, where we describe the build and deploy part of the workflow. This file is generated using the `skaffold init` command. We will explore this and many other Skaffold CLI commands in *Chapter 5, Installing Skaffold and Demystifying Its Pipeline Stages*. Skaffold typically expects the `skaffold.yaml` configuration file in the current directory, but you can override it by passing the `--filename` flag.

This is the content of the configuration file:

```
apiVersion: skaffold/v2beta20
kind: Config
metadata:
  name: indian-states
build:
  artifacts:
    - image: docker.io/hiashish/skaffold-introduction
      jib: {}
deploy:
  kubectl:
    manifests:
      - k8s/mydeployment.yaml
      - k8s/myservice.yaml
```

Let me just explain the key components in this file, as follows:

- `apiVersion`: This specifies the **application programming interface (API)** schema version.
- `build`: This specifies how images are built with Skaffold.
- `artifacts`: Here, we have the images to be built.
- `image`: This is the name of the image to be built.
- `jib`: This specifies that the image is built using the Jib Maven plugin.
- `deploy`: This specifies how the image is going to be deployed to a local or remote Kubernetes cluster.
- `kubectl`: This specifies that the `kubectl` CLI is going to be used for creating and updating Kubernetes manifests.
- `manifests`: This specifies the Kubernetes manifest file path—that is, deployments and services.

Now you have understood the Skaffold configuration, the next logical step is to build and deploy our Spring Boot application using Skaffold.

Building and deploying the Spring Boot application

Before we go ahead with the build and deployment of our Spring Boot application, please make sure that Docker is up and running before running the `skaffold` command. Otherwise, you will get the following error:

```
Cannot connect to the Docker daemon at unix:///var/run/docker.
sock. Is the docker daemon running?
```

The only thing now remaining is to run the `skaffold dev` command and start the **continuous development (CD)** workflow. If you run this command without enabling Kubernetes with Docker Desktop, it will fail, with the following error. So, watch out for these prerequisites:

```
Deploy Failed. Could not connect to cluster docker-
desktop due to "https://kubernetes.docker.internal:6443/
version?timeout=32s": dial tcp 127.0.0.1:6443: connect:
connection refused. Check your connection for the cluster.
```

If all the prerequisites are met, then the moment you enter that command, what Skaffold will do is watch for changes in your source code directory using its **File Watcher** mechanism. It will build an image, push it to the local Docker registry, deploy your application, and stream logs from the running pods.

How cool is that?!! You should see the following output:

```
$ skaffold dev
Listing files to watch...
- docker.io/hiashish/skaffold-introduction
Generating tags...
- docker.io/hiashish/skaffold-introduction -> docker.io/
hiashish/skaffold-introduction:22f18cc-dirty
Checking cache...
- docker.io/hiashish/skaffold-introduction: Not found. Building
Starting build...
Found [docker-desktop] context, using local docker daemon.
Building [docker.io/hiashish/skaffold-introduction]...
[INFO] --- jib-maven-plugin:2.8.0:dockerBuild (default-cli) @
skaffold-introduction ---
[INFO] Containerizing application to Docker daemon as hiashish/
skaffold-introduction:22f18cc-dirty...
[WARNING] Base image 'openjdk:16-jdk-alpine' does not use a
specific image digest - build may not be reproducible
[INFO] Building dependencies layer...
[INFO] Building resources layer...
[INFO] Building classes layer...
[INFO] The base image requires auth. Trying again for
openjdk:16-jdk-alpine...
[INFO] Using credentials from Docker config (/Users/ashish/.
docker/config.json) for openjdk:16-jdk-alpine
[INFO] Using base image with digest: sha256:49d822f4fa4deb
5f9d0201ffeec9f4d113bcb4e7e49bd6bc063d3ba93aacbcae
[INFO] Container entrypoint set to [java, -cp, /app/
resources:/app/classes:/app/libs/*, com.example.indianstates.
IndianStatesApplication]
[INFO] Loading to Docker daemon...
[INFO] Built image to Docker daemon as hiashish/skaffold-
introduction:22f18cc-dirty
[INFO] BUILD SUCCESS
```

> **Note**
> To reduce the verbosity of the logs, we have trimmed them down to show only the parts that are relevant to our discussion.

Since there are lots of logs generated and it would be difficult to explain them all at once, I have intentionally kept these in chunks to help you understand the Skaffold working better through these logs. So far, we can conclude the following from the logs:

- Skaffold first tries to figure out the source code dependencies it needs to watch based on the builder defined in the `skaffold.yaml` file.

- It then generates a tag for the image, as mentioned in the `build` section of the `skaffold.yaml` file. You might be wondering why an image tag is generated before the image is built. We will cover the Skaffold tagging mechanism specifically in *Chapter 5, Installing Skaffold and Demystifying Its Pipeline Stages*.

- Then, it tries to find the image in the local cache. Images are cached locally to improve the execution time if there is no need for compilation. Since the image was not available locally, Skaffold started the build.

Before doing the actual build, Skaffold identified that the Kubernetes context is set to `docker-desktop`. It will use a local Docker daemon to create an image. Did you see the clever guesses it takes to fasten the inner development loop? You can verify the current `kube-context` status with the following command:

```
$kubectl config current-context
docker-desktop
```

Since we are using the `jib-maven-plugin` plugin and the Kubernetes context is set to `docker-desktop`, Skaffold will internally use the `jib:dockerBuild` command to create an image. We have used `openjdk:16-jdk-alpine` as the base image because it's lightweight.

First, Jib will try to authenticate with the Docker Hub container registry and download the base image using credentials from the `config.json` file located at the `/Users/ashish/.docker/config.json` path; then, it will create image layers, and finally upload it to the local Docker daemon, as seen in the following example:

```
Starting test...
Tags used in deployment:
- docker.io/hiashish/skaffold-introduction -> docker.io/
hiashish/skaffold-introduction:adb6df1b0757245bd08f790e93ed5f8c
c621a8f7e500e3c5ad18505a8b677139
```

```
Starting deploy...
- deployment.apps/skaffold-introduction created
- service/skaffold-introduction created
Waiting for deployments to stabilize...
- deployment/skaffold-introduction is ready.
Deployments stabilized in 3.771 seconds
Press Ctrl+C to exit
Watching for changes...
[skaffold-introduction]   :: Spring Boot
::                 (v2.4.4)
[skaffold-introduction] 2021-03-25 21:17:49.048   INFO 1 ---
[          main] c.e.i.IndianStatesApplication              :
Starting IndianStatesApplication using Java 16-ea on skaffold-
introduction-85bbfddbc9-bfxnx with PID 1 (/app/classes started
by root in /)
[skaffold-introduction] 2021-03-25 21:17:55.895   INFO 1 --- [
main] o.s.b.w.embedded.tomcat.TomcatWebServer   : Tomcat started
on port(s): 8080 (http) with context path ''
[skaffold-introduction] 2021-03-25 21:17:55.936   INFO 1 ---
[          main] c.e.i.IndianStatesApplication              :
Started IndianStatesApplication in 8.315 seconds (JVM running
for 9.579)
```

We can conclude the following from the logs:

- In the first line, in the `Starting test...` logs, Skaffold runs container-structure tests to validate built container images before deploying them to our cluster.

- Soon after that, Skaffold will create Kubernetes manifests—that is, deployment and services available under the `k8s` directory.

- Once the manifests are created, it means the pod is up and running after some time. Then, it will also start streaming logs from the pod on your console itself.

We will now do some verification to make sure that the pod is actually running. We will run the following `kubectl` command for verification:

```
NAME                                            READY   STATUS    RESTARTS   AGE
pod/skaffold-introduction-667786cc47-khx4q      1/1     Running   0          3m11s

NAME                            TYPE        CLUSTER-IP       EXTERNAL-IP   PORT(S)          AGE
service/kubernetes              ClusterIP   10.96.0.1        <none>        443/TCP          30h
service/skaffold-introduction   NodePort    10.105.226.176   <none>        8080:30368/TCP   3m11s

NAME                                      READY   UP-TO-DATE   AVAILABLE   AGE
deployment.apps/skaffold-introduction     1/1     1            1           3m11s

NAME                                               DESIRED   CURRENT   READY   AGE
replicaset.apps/skaffold-introduction-667786cc47   1         1         1       3m11s
```

Figure 3.5 – Kubernetes resources created

As you can see, we have a pod named `skaffold-introduction-667786cc47-khx4q` with a RUNNING status. Let's hit the `/states` REST endpoint and see if we are getting the desired output or not, as follows:

```
$ curl localhost:30368/states
[{"name":"Andra Pradesh","capital":"Hyderabad"},{"name":"Arunachal Pradesh",
"capital":"Itangar"},{"name":"Assam","capital":"Dispur"},
{"name":"Bihar","capital":"Patna"},{"name":"Chhattisgarh",
"capital":"Raipur"},{"name":"Goa","capital":"Panaji"},
{"name":"Gujarat","capital":"Gandhinagar"},{"name":"Haryana",
"capital":"Chandigarh"},{"name":"Himachal
Pradesh","capital":"Shimla"},{"name":"Jharkhand",
"capital":"Ranchi"},{"name":"Karnataka",
"capital":"Bengaluru"},{"name":"Kerala","capital":
"Thiruvananthapuram"},{"name":"Madhya
Pradesh","capital":"Bhopal"},{"name":"Maharashtra",
"capital":"Mumbai"},{"name":"Manipur","capital":"Imphal"},
{"name":"Meghalaya","capital":"Shillong"},{"name":"Mizoram",
"capital":"Aizawl"},{"name":"Nagaland","capital":"Kohima"},
{"name":"Orissa","capital":"Bhubaneshwar"},
{"name":"Rajasthan","capital":"Jaipur"},
{"name":"Sikkim","capital":"Gangtok"},
{"name":"Tamil Nadu","capital":"Chennai"},{"name":"Telangana",
"capital":"Hyderabad"},{"name":"Tripura","capital":"Agartala"},
{"name":"Uttarakhand","capital":"Dehradun"},
{"name":"Uttar Pradesh","capital":"Lucknow"},
{"name":"West Bengal","capital":"Kolkata"},
{"name":"Punjab","capital":"Chandigarh"}]
```

Indeed we are getting the expected output. Let's hit the other
/state?name=statename REST endpoint as well and see if we are getting
the desired output or not, as follows:

```
$ curl -X GET "localhost:30368/state?name=Karnataka"
Bengaluru
```

Yes—we do get the desired output!

When you run the skaffold dev command, it creates a CD pipeline. For example, if
there are any code changes in this mode, Skaffold will automatically rebuild and redeploy
the image.

In Skaffold dev mode, since we are using a local Kubernetes cluster and the
Kubernetes context is set to docker-desktop, Skaffold will not push the image to a
remote container registry and will load it to a local Docker registry instead. It will further
help in accelerating the inner development loop.

Finally, to clean up everything we have done so far, we can just press *Ctrl + C*, and Skaffold
will take care of the rest.

We thus reach the end of this demonstration, where we have successfully built and
deployed a Spring Boot application to a single-node Kubernetes cluster that comes with
Docker Desktop, using Skaffold.

Summary

In this chapter, we introduced you to Skaffold and some of its commands and concepts. In
the example, we have introduced you to only one Skaffold command—that is, skaffold
dev. However, there are many such commands, for example, skaffold run and
skaffold render, which we will cover in upcoming chapters. You have also learned
how to build and deploy applications with Skaffold using commands such as skaffold
dev to the local Kubernetes cluster.

In the next chapter, we will learn about Skaffold features and architectures.

Further reading

- Learn more about developing Java applications with Spring Boot from *Developing
  Java Applications with Spring and Spring Boot*, published by Packt Publishing
  (https://www.packtpub.com/product/developing-java-
  applications-with-spring-and-spring-boot/9781789534757).

- The most recent Java 16 release notes are available at https://jdk.java.
  net/16/.

Section 2: Getting Started with Skaffold

In this section, we will cover Skaffold features and internal architecture. We will try to understand some diagrams depicting how Skaffold works. We will also learn how to bootstrap our project with Skaffold by covering some basics of the Skaffold configuration file. We will learn about Skaffold installation and various commands that we can use in different pipeline stages, that is, `init`, `build`, and `deploy`. Then, finally, in this section, we explain different ways of building and deploying container images with Skaffold.

In this section, we have the following chapters:

- *Chapter 4, Understanding Skaffold's Features and Architecture*
- *Chapter 5, Installing Skaffold and Demystifying Its Pipeline Stages*
- *Chapter 6, Working with Skaffold Container Image Builders and Deployers*

4
Understanding Skaffold's Features and Architecture

In the previous chapter, we gained a basic understanding of Skaffold through some coding examples. This chapter will cover the features that are provided by Skaffold. Additionally, we will explore Skaffold's internals by looking at its architecture, workflow, and the `skaffold.yaml` configuration file.

In this chapter, we're going to cover the following main topics:

- Understanding Skaffold's features
- Demystifying Skaffold's architecture
- Understanding the Skaffold workflow
- Deciphering Skaffold's configuration with `skaffold.yaml`

By the end of this chapter, you will have a solid understanding of the features that Skaffold provides, and how it performs all the magic, by looking at its workflow and architecture.

Technical requirements

To follow along with the examples in this chapter, you will need to have the following software installed:

- Eclipse (`https://www.eclipse.org/downloads/`) or IntelliJ IDE (`https://www.jetbrains.com/idea/download/`)

- Git (`https://git-scm.com/downloads`)

- Skaffold (`https://skaffold.dev/docs/install/`)

- Docker Desktop for macOS and Windows (`https://www.docker.com/products/docker-desktop`)

You can download the code examples for this chapter from the GitHub repository at `https://github.com/PacktPublishing/Effortless-Cloud-Native-App-Development-Using-Skaffold`.

Understanding Skaffold's features

In *Chapter 3, Skaffold – Easy-Peasy Cloud-Native Kubernetes Application Development*, we were introduced to Skaffold. We uncovered some of its features by building and deploying a Spring Boot application to the local Kubernetes cluster. However, Skaffold is capable of far more than that, so let's take a look at some of its features.

Skaffold has the following features:

- **Easy to share**: Sharing your project among the same or different teams, provided they already have Skaffold installed, is effortless. They have to run the following commands to proceed with their development activities:

```
git clone repository URL
skaffold dev
```

- **Integrated with IDE**: Many IDEs, such as IntelliJ and VS Code, support the **Cloud Code** plugin developed by Google, which internally uses Skaffold and its API to provide a better developer experience while developing a Kubernetes application. Using the IntelliJ or VS code **Google Cloud Code Extension** plugin makes it easier for you to create, edit, and update the `skaffold.yaml` file with its code completion feature. For example, to give you a little more context regarding this, the plugin can detect that the project is using Skaffold to build and deploy by looking at the `skaffold.yaml` configuration file:

Figure 4.1 – The IntelliJ Cloud code plugin detects the Skaffold configuration

You can also look up supported builders and deployers with Skaffold using the code completion feature. We will cover the Cloud Code plugin specifically in *Chapter 7, Building and Deploying a Spring Boot Application with the Cloud Code Plugin.*

- **File sync**: Skaffold has an excellent file sync feature. It can copy the changed files directly to an already running container to avoid rebuilding, redeploying, and restarting the container.

  We will learn more about this in *Chapter 5, Installing Skaffold and Demystifying Its Pipeline Stages.*

- **Super-fast local development**: In the previous chapter, you learned that building and deploying an application with Skaffold is pretty quick as it can figure out whether your Kubernetes context is set to a local Kubernetes cluster and will avoid pushing the image to a remote container registry. Therefore, you can bypass that expensive network hop and also preserve the battery life of your laptop.

  Not only that, but Skaffold detects your code changes in real time and automates the build, push, and deploy workflows. So, you can continue working within your inner development loop with the focus on coding, and there is no need to leave that loop until you are completely sure about the change you are making. This not only accelerates your inner development loop but also makes you more productive.

- **Effortless remote development**: So far, while reading this book, you might have the impression that Skaffold can only accelerate the inner development loop. Oh, boy! You are in for a surprise as Skaffold can handle outer development loop workflows as well. For example, you can use Skaffold to create full-fledged production-ready CI/CD pipelines. We will cover this specifically in *Chapter 9, Creating a Production-Ready CI/CD Pipeline with Skaffold*. Not only that, but you can also switch your Kubernetes context in your local development environment using commands such as `kubectl config use-context context-name` and perform a deployment to a remote cluster of your choice.

Since we are talking about remote development, I just wanted to highlight another point—you don't need to run Docker daemon if you are using the `jib-maven` plugin for a remote build (that is, if you are pushing to a remote container registry). You can also use something like **Google Cloud Build** to do remote builds. Cloud Build is a service provided by **Google Cloud Platform** that you can use to execute your builds and create serverless CI/CD pipelines in the cloud for your cloud-native applications. It might be slow if you run it from your local system, but it's worth exploring.

- **Built-in image tag management**: In the previous chapter, while declaring Kubernetes deployment manifests, we only mentioned the image name and not the image tag while building and deploying the Spring Boot application. For example, in the following snippet from the previous chapter, in the `image:` field, we only mentioned the image name:

```
spec:
    containers:
        - image: docker.io/hiashish/skaffold-introduction
          name: skaffold-introduction
```

Typically, we have to tag an image before pushing and then use the same image tag while pulling. For example, you would also have to specify the image tag in the following format:

```
- image: imagename:imagetag
```

The reason for this is that Skaffold automatically generates the image tag for you, out of the box, every time you rebuild the image. This is so that you don't have to edit the Kubernetes manifests manually. The default tagging strategy with Skaffold is `gitCommit`.

We will cover this in greater detail in *Chapter 5, Installing Skaffold and Demystifying Its Pipeline Stages*.

- **Lightweight**: Skaffold is entirely a CLI tool. There are no server-side components to look for while working with Skaffold. This makes it extremely lightweight, easy to use, and there is no maintenance burden. The size of the Skaffold binary is around 63 MB.

- **Pluggable architecture**: Skaffold has a pluggable architecture. This will eventually mean that you can pick and choose a build and deploy tool of your choice. Bring your own tools, and Skaffold will adjust itself accordingly.

- **Purpose-built for CI/CD pipelines**: Skaffold can help you to create effective CI/CD pipelines. For example, you can use the `skaffold run` command to execute an end-to-end pipeline or use individual commands such as `skaffold build` or `skaffold deploy`.

  Furthermore, with commands such as `skaffold render` and `skaffold apply`, you can create a **GitOps**-style continuous delivery workflow for your application. GitOps allows you to store your application's desired state inside a Git repository in the form of Kubernetes manifests. It also allows others to view your infrastructure as code.

- **Effortless environment management**: Skaffold allows you to define, build, test, and deploy configurations for different environments. You can keep one set of configurations for development or staging and another for production. Also, you can keep completely different configurations on a per-environment basis. You can achieve this by using Skaffold profiles. This is relatively similar to the `profiles` feature provided for Spring Boot applications.

  Please refer to the following screenshot:

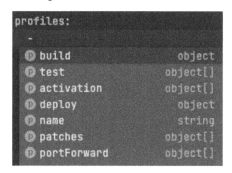

Figure 4.2 – Skaffold profiles

A typical Skaffold profile consists of the following parts:

- `build`
- `test`
- `activation`
- `deploy`
- `name`
- `patches`

Some of these parts are quite obvious, as they explain the unique name of the profile, the build steps, the deployment steps, and how the images are tested. Let's move on to discuss patches and activation.

First, let's understand patches.

Skaffold profile patches

As the name suggests, patches are a more verbose way of overriding individual values in a skaffold.yaml file. For example, in the following snippet, instead of overriding the whole build section, the dev profile defines a different Dockerfile for the first artifact:

```
build:
  artifacts:
    - image: docker.io/hiashish/skaffold-example
      docker:
        dockerfile: Dockerfile
    - image: docker.io/hiashish/skaffold2
    - image: docker.io/hiashish/skaffold3
deploy:
  kubectl:
    manifests:
      - k8s-pod
profiles:
  - name: dev
    patches:
      - op: replace
        path: /build/artifacts/0/docker/dockerfile
        value: Dockerfile_dev
```

Here, the op string underneath the patches section specifies the operation to be performed by this patch. The path string specifies the position in the .yaml file where the operation that you defined in the op string takes place, and the value object specifies the value it should be replaced with.

The following operations are supported:

- add
- remove
- replace

- `move`
- `copy`
- `test`

To summarize, here, we instruct Skaffold to replace the `Dockerfile` that was used to build the first `docker.io/hiashish/skaffold-example` image with a different `Dockerfile` named `Dockerfile_dev`.

Now, let's discuss activation objects in profiles.

Skaffold profile activation

You can activate a profile in Skaffold in one of the following two ways:

- Using a CLI
- With `skaffold.yaml` activations

First, let's discuss how you can activate a profile using CLI. For example, in the following `skaffold.yaml` file, underneath the `profiles` section, we have declared a profile name, called `gcb`:

```
apiVersion: skaffold/v2beta20
kind: Config
metadata:
  name: skaffold-introduction
build:
  artifacts:
    - image: docker.io/hiashish/skaffold-introduction
      jib: { }
deploy:
  kubectl:
    manifests:
      - k8s/mydeployment.yaml
      - k8s/myservice.yaml
profiles:
  - name: gcb
    build:
      googleCloudBuild:
        projectId: gke_projectid
```

This profile will be activated when running the `skaffold run` or `skaffold dev` command by passing either the `--profile` or `-p` CLI flag. If you run the following command, then Skaffold will use **Google Cloud Build** to build the artifacts:

```
skaffold run -p gcb
```

Notice that we have not specified the `deploy` section underneath the `gcb` profile. This means that Skaffold will continue to use `kubectl` for deployment. If your use case requires multiple profiles, you can use the `-p` flag numerous times or pass comma-separated profiles, as shown in the following command:

```
skaffold dev -p profile1,profile2
```

Let's try to understand this using another example. In this example, we will use the Spring Boot application that we built in *Chapter 3, Skaffold – Easy-Peasy Cloud-Native Kubernetes Application Development*. In that scenario, we used Jib to containerize the application; however, in this example, we will use a multistage Docker build to create a slim Docker image of our application. The following is the `Dockerfile` of our Spring Boot application:

```
FROM maven:3-adoptopenjdk-16 as build
RUN mkdir /app
COPY . /app
WORKDIR /app
RUN mvn clean package

FROM adoptopenjdk:16-jre
RUN mkdir /project
COPY --from=build /app/target/*.jar /project/app.jar
WORKDIR /project
ENTRYPOINT ["java","-jar","app.jar"]
```

We can explain the multistage `Dockerfile` build as follows:

- In the first stage of the build, we have used the `maven:3-adoptopenjdk-16` image to build and create `jar` for our application using the `mvn clean package` Maven command.
- In the second stage, we have copied `jar` we made in the previous build stage and created a new final image based on a significantly smaller *Java 16 JRE base image*.

- The final Docker image does not include the JDK or Maven image, just the JRE image. The only downside to this approach is that the build time is higher because all of the required dependencies need to be downloaded during the first stage of the build.

> **Tip**
>
> You can use Docker multistage builds to create slimmer Docker images of your application. The size of a typical JDK image is around 650 MB, and with the Docker multistage build, we can reduce its size by half by using JRE as the base image in the last stage of the multistage build.
>
> Additionally, you can further reduce the size of the image using Java tools such as jdeps and jlink (introduced in Java 9). While jdeps helps you to identify the required JVM modules, jlink allows you to create a customized JRE. With the combination of these tools, you can create a customized JRE, which results in an even slimmer Docker image of your application.

To demonstrate the use of profiles, we will make the following changes to the skaffold. yaml file. The following is the skaffold.yaml file in which we added a new profile called docker:

```
apiVersion: skaffold/v2beta20
kind: Config
metadata:
  name: skaffold-introduction
build:
  artifacts:
    - image: docker.io/hiashish/skaffold-introduction
      jib: { }
deploy:
  kubectl:
    manifests:
      - k8s/mydeployment.yaml
      - k8s/myservice.yaml
profiles:
  - name: docker
    build:
      artifacts:
        - image: docker.io/hiashish/skaffold-introduction
```

```
docker:
    dockerfile: Dockerfile
```

We will use the skaffold run --profile docker command to build and deploy our Spring Boot application. The following is the output:

```
Generating tags...
- docker.io/hiashish/skaffold-introduction -> docker.io/
hiashish/skaffold-introduction:fcda757-dirty
Checking cache...
- docker.io/hiashish/skaffold-introduction: Not found. Building
Starting build...
Found [minikube] context, using local docker daemon.
Building [docker.io/hiashish/skaffold-introduction]...
Sending build context to Docker daemon  128.5kB
Step 1/10 : FROM maven:3-adoptopenjdk-16 as build
3-adoptopenjdk-16: Pulling from library/maven
........
ecf4fc483ced: Pull complete
Status: Downloaded newer image for maven:3-adoptopenjdk-16
---> 8bb5929b61c3
Step 2/10 : RUN mkdir /app
---> Running in ff5bf71356dc
---> 83040b88c925
Step 3/10 : COPY . /app
---> 5715636b31d8
Step 4/10 : WORKDIR /app
---> Running in 6de38bef1b56
---> ca82b0631625
Step 5/10 : RUN mvn clean package -DskipTests
---> Running in 91df70ce44fa
[INFO] Scanning for projects...
Downloading from repository.spring.milestone: https://repo.
spring.io/milestone/org/springframework/boot/spring-boot-
starter-parent/2.5.0-M1/spring-boot-starter-parent-2.5.0-M1.pom
........
[INFO] BUILD SUCCESS
```

In the preceding logs, you can see that, first, Skaffold began building our image using Docker. Also, we have used a multistage build, and then, in steps 1 to 6, we are into the first stage of the build, wherein we created the jar of our application inside the container:

```
Step 6/10 : FROM adoptopenjdk:16-jre
16-jre: Pulling from library/adoptopenjdk
c549ccf8d472: Already exists
........
23bb7f46497d: Pull complete
Digest: sha256:f2d0e6433fa7d172e312bad9d7b46ff227888926f2fe526
c731dd4de295ef887
Status: Downloaded newer image for adoptopenjdk:16-jre
---> 954409133efc
Step 7/10 : RUN mkdir /project
---> Running in abfd14b21ac6
---> 2ab11f2093a3
Step 8/10 : COPY --from=build /app/target/*.jar /project/app.
jar
---> 52b596edfac9
Step 9/10 : WORKDIR /project
---> Running in 473cbb6d878d
---> b06856859039
Step 10/10 : ENTRYPOINT ["java","-jar","app.jar"]
---> Running in 6b22aee242d2
---> f62822733ebd
Successfully built f62822733ebd
Successfully tagged hiashish/skaffold-introduction:fcda757-
dirty
```

in *steps 6 to 10*, we are in the second stage of the build where we have used adoptopenjdk:16-jre as the base image since we only need JRE to run our application. Typically, JRE images are of a smaller size than JDK.

This final output is our containerized application, which should appear as follows:

```
Starting test...
Tags used in deployment:
- docker.io/hiashish/skaffold-introduction -> docker.io/
hiashish/skaffold-introduction:f62822733ebd832cab
```

```
058e5b0282af6bb504f60be892eb074f980132e3630d88
Starting deploy...
- deployment.apps/skaffold-introduction created
- service/skaffold-introduction created
Waiting for deployments to stabilize...
- deployment/skaffold-introduction is ready.
Deployments stabilized in 4.378 seconds
```

Finally, Skaffold deploys our containerized application to the local Kubernetes cluster.

Another way to activate a profile is to use the activation object array in `skaffold.yaml` to automatically activate a profile based upon the following:

- `kubeContext`
- An environment variable: `env`
- A Skaffold command

Please refer to the following screenshot:

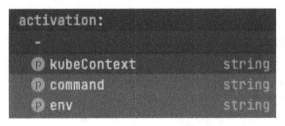

Figure 4.3 – Skaffold profile activation using the skaffold.yaml file

Let's attempt to understand this activation option using an example.

In the following code example, we have two profiles—`profile-staging` and `profile-production`. As their names suggest, `profile-staging` will be used for the staging environment, while `profile-production` will be used for the production environment:

```
build:
  artifacts:
    - image: docker.io/hiashish/skaffold-introduction
      jib: { }
deploy:
  kubectl:
    manifests:
```

```
        - k8s/mydeployment.yaml
        - k8s/myservice.yaml
 profiles:
   - name: profile-staging
     activation:
        - env: ENV=staging
   - name: profile-production
     build:
       googleCloudBuild:
         projectId: gke_projectid
     activation:
        - env: ENV=production
        - kubeContext: gke_cluster
          command: run
```

Here, `profile-staging` will be automatically activated if the `ENV` environment variable key matches the value staging. We have not specified the build, test, and deploy steps for this specific profile, so it will continue to use the options we have provided in the main section of the `skaffold.yaml` file. In addition to this, `profile-production` will be automatically activated if the following criteria are satisfied. Note that it only runs the profile production stage if all of these criteria are activated:

- The `ENV` environment variable key matches the value production.

- The Kubernetes context is set to **GKE** (which is short for **Google Kubernetes Engine**).

- The Skaffold command that is used is `scaffold run`.

Note that `profile-production` will use Google's Cloud Build for the build and will default to `kubectl` for the deployment (as it's not explicitly specified).

This segregation also allows you to use various tools to build and deploy within different environments. For example, you might use Docker to create images in a local development and `Jib` for production. In the case of a deployment, you might use `kubectl` in development and Helm in production.

In the previous chapter, I explained that Skaffold, by default, looks for the current Kubernetes context from the `kube config` file that is located in `${HOME}/.kube/config path`. If you wish to change it, you can do that while running the `skaffold dev` command:

```
skaffold dev --kube-context <myrepo>
```

You can also mention `kubeContext` in the `skaffold.yaml` file as follows:

```
deploy:
  kubeContext: docker-desktop
```

The flag passed via the CLI takes precedence over the `skaffold.yaml` file.

Next, let's discuss how Skaffold configures or adjusts itself to different local Kubernetes clusters.

A local Kubernetes cluster

By now, you should have realized that Skaffold provides sensible, smart defaults to make the development process easier without you having to tell it to do things. If your Kubernetes context is set to a local Kubernetes cluster, then there is no need to push an image to a remote Kubernetes cluster. Instead, Skaffold will move the image to the local Docker daemon to speed up the development cycle.

So far, we have only discussed the Kubernetes cluster that comes with Docker Desktop, but this is not the only option you have. There are various ways in which to set up and run a local Kubernetes cluster. For example, you have the following choices when creating a local Kubernetes cluster:

- Docker Desktop (`https://docs.docker.com/desktop/kubernetes/#enable-kubernetes`)
- Minikube (`https://minikube.sigs.k8s.io/docs/`)
- Kind (`https://kind.sigs.k8s.io/`)
- k3d (`https://k3d.io/`)

If any of these supported Kubernetes installations are available for local development, Skaffold expects that Kubernetes' context is as shown in the following table. Otherwise, it will assume that we are deploying to a remote Kubernetes cluster.

Skaffold detects a local cluster based on the Kubernetes context name, as described in the following table:

Kubernetes context	Local cluster	Remarks
`docker-desktop`	Docker Desktop (`https://www.docker.com/products/docker-desktop`)	
`docker-for-desktop`	Docker Desktop (`https://www.docker.com/products/docker-desktop`)	This context is deprecated.
`Minikube`	minikube (`https://github.com/kubernetes/minikube/`)	
`kind-(.*)`	kind (`https://github.com/kubernetes-sigs/kind`)	This pattern is used if the kind version is ≥ v0.6.0.
`(.*)@kind`	kind (`https://github.com/kubernetes-sigs/kind`)	This pattern is used if the kind version is < v0.6.0
`k3d-(.*)`	k3d (`https://github.com/rancher/k3d`)	This pattern is used if the k3d version is ≥ v3.0.0

Table 4.1 – The Kubernetes context supported by Skaffold

However, for other nonstandard local cluster setups, such as when running `minikube` with a custom profile (for instance, `minikube start -p my-profile`), you can use the following commands to tell Skaffold that you are using a local Kubernetes cluster:

1. First, set up the Docker environment for Skaffold using the following command:

    ```
    source <(minikube docker-env -p my-profile)
    ```

2. Then, instruct Skaffold to consider `my-profile` as a local cluster using the following command:

    ```
    $ skaffold config set --kube-context my-profile local-cluster true
    ```

In this section, we took a deep dive into the features that Skaffold provides. Now, let's discuss Skaffold's architecture.

Demystifying Skaffold's architecture

As mentioned in the previous section, Skaffold has been designed with pluggability in mind. The following is a visualization of the Skaffold architecture:

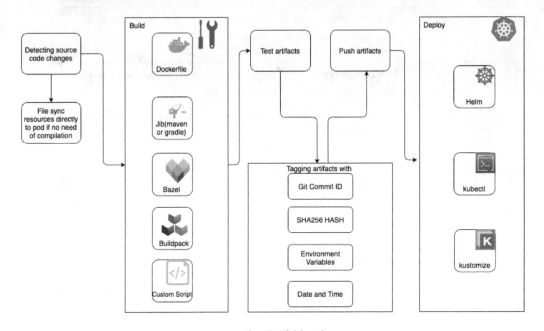

Figure 4.4 – The Skaffold architecture

From this architecture diagram, you can conclude that Skaffold has a modular design. However, what is a modular design?

Well, a modular design, or modularity in design, is a design principle that subdivides a system into smaller parts called modules, which can be independently created, modified, replaced, or exchanged with other modules or between different systems.

With this definition, we can define the following modules for Skaffold:

- Container image builders
- Container testing tools/strategy
- Container image taggers
- Container deployment tools

Now, let's discuss each of these tools/modules in more detail. Currently, Skaffold supports the following container image builders:

- **Dockerfile**
- **Jib (Maven and Gradle)**
- **Bazel**
- **Cloud-Native Buildpacks**
- **Custom Script**

For deployment to Kubernetes, Skaffold supports the following tools:

- **Helm**
- **kubectl**
- **kustomize**

We will discuss these options in greater detail in *Chapter 6, Working with Skaffold Container Image Builders and Deployers.*

Skaffold supports the following types of tests between the build and deployment phases of the pipeline:

- Custom tests
- Container structure tests

We will explore these options further in *Chapter 5, Installing Skaffold and Demystifying Its Pipeline Stages.*

As mentioned earlier, underneath the Skaffold **Features** section, Skaffold offers built-in image tag management. Currently, Skaffold supports multiple taggers and tag policies to tag images:

- The `gitCommit` tagger
- The `inputDigest` tagger
- The `envTemplate` tagger
- The `datetime` tagger
- The `customTemplate` tagger
- The `sha256` tagger

Knowing which image tag policy is supported is very easy with the IntelliJ Cloud Code plugin code completion feature. Let's suppose you don't specify the image tag policy in the `skaffold.yaml` file; in that case, the default policy is the `gitCommit` tagger:

Take a look at the following screenshot:

Figure 4.5 – Skaffold's supported image tag policies

Now, considering the pluggable architecture of Skaffold, you might use **Local Docker Daemon** for building images, `kubectl` for deployment to `minikube`, or any other supported local Kubernetes cluster. In this scenario, Skaffold will not push the image to a remote registry, and you can even skip the container structure tests by using the `-skipTests` flag.

The following diagram shows the tools used for local development in this scenario:

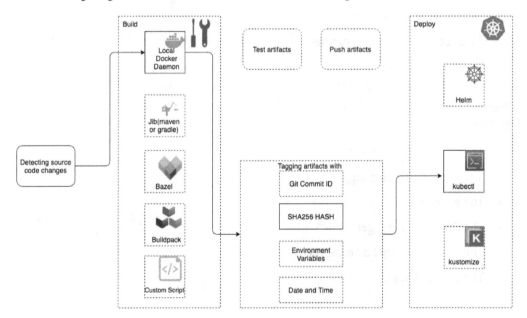

Figure 4.6 – Skaffold in development

While, in the case of the production scenario, you might use a Jib Maven or Gradle plugin to build the image, test the artifacts, push it to the remote registry, and, finally, deploy it to the remote Kubernetes cluster using Helm.

The following diagram shows the tools used for the production scenario:

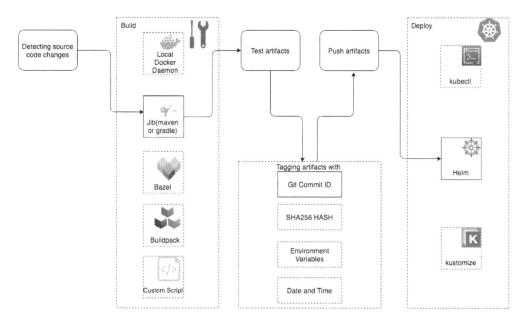

Figure 4.7 – Skaffold in production

This completes our deep-dive analysis of the Skaffold architecture. Now, let's discuss the Skaffold workflow.

Understanding the Skaffold workflow

Typically, Skaffold works in two modes, namely, *continuous development* or an *end-to-end pipeline* through commands such as skaffold dev and skaffold run. For example, when you run the skaffold dev command, the following steps are carried out by Skaffold:

1. Receive and watch your source code for changes.
2. Copy the changed files straight to build if the user marks them as eligible for copying.
3. Build your artifacts from the source code.
4. Test your built artifacts using container-structure-tests or custom scripts.

5. Tag your artifacts.

6. Push your artifacts (only if the Kubernetes context is set to a remote cluster).

7. Update the Kubernetes manifests with the correct tags.

8. Deploy your artifacts.

9. Monitor the deployed artifacts with built-in health checks.

10. Stream logs from the running pods.

11. Clean up any deployed artifacts on exit by pressing *Ctrl + C*.

    In the case of the `skaffold run` command, the workflow is relatively similar. The only difference is the following:

- Skaffold will not continuously watch for code changes.

- By default, Skaffold will not stream logs from the running pods.

- Skaffold will exit after *step 9* in the end-to-end pipeline mode.

The following diagram illustrates both the continuous development and end-to-end pipeline that we explained in the preceding steps:

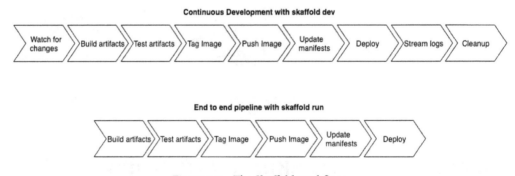

Figure 4.8 – The Skaffold workflow

You should now have an understanding of how Skaffold works in continuous development and end-to-end pipeline mode. Let's take a look at the components that are available in its configuration with the `skaffold.yaml` file.

Deciphering Skaffold's configuration with skaffold.yaml

Any action that Skaffold needs to perform should be clearly defined in the `skaffold.yaml` configuration file. In this configuration file, you must specify which tool Skaffold has to use to build an image and then deploy it to the Kubernetes cluster. Skaffold typically expects to find the configuration file as `skaffold.yaml` in the current directory; however, we can override the location using the `--filename` flag.

> **Tip**
> We recommend that you keep the Skaffold configuration file in the root directory of the project.

The configuration file consists of the following main components:

Component	Description
apiVersion	This defines the Skaffold API version that you would like to use. At the time of writing this book, the current API version is `skaffold/v2beta20`.
Kind	The Skaffold configuration file uses the kind configuration.
Metadata	This contains additional properties such as the name of this configuration.
Build	This defines how Skaffold will build the artifacts. You have the freedom to decide what tools Skaffold can use, such as Docker, Kaniko, or Buildpacks. In addition to this, you can also define how Skaffold tags or pushes artifacts.
Test	This defines how Skaffold will test artifacts. Skaffold supports `container-structure-tests` (`https://github.com/GoogleContainerTools/container-structure-test`) and custom tests to test built artifacts.
Deploy	This defines how Skaffold will deploy the artifacts using tools such as kubectl, helm, or, kustomize.
profiles	This defines a set of settings that, when activated, allows you to override the current configuration, such as the build, test, and deploy sections.
Requires	This defines a list of additional Skaffold configurations to import into the current configuration.

Table 4.2 – The skaffold.yaml file components

Skaffold also supports a global configuration file, which is located in the `~/.skaffold/config` path. The following are the options it supports, which can be defined at the global level:

Option	Type	Description
default-repo	String	This specifies the image registry where built artifact images are pushed.
debug-helpers-registry	String	This specifies the image registry where debug support images are retrieved.
insecure-registries	List of strings	This specifies a list of image registries that might be accessed without TLS.
k3d-disable-load	Boolean	If this flag is set to true, then do not use the k3d import image to load images locally.
kind-disable-load	Boolean	If this flag is set to true, then do not use the kind load to load images locally.
local-cluster	Boolean	If this flag is set to true, then do not try to push images after building. By default, contexts with the names of docker-desktop or minikube are treated as local.

Table 4.3 – Skaffold global configuration options

You can list, set, and unset these options in the command line easily by using the following commands:

```
$ skaffold config
Interact with the Skaffold configuration
Available Commands:
  list       List all values set in the global Skaffold config
  set        Set a value in the global Skaffold config
  unset      Unset a value in the global Skaffold config
```

For example, you can set the local cluster option to false. This will allow you to push an image to a remote registry after building the image. Please refer to the following commands:

```
$ skaffold config set --global local-cluster false
set global value local-cluster to false
$ cat ~/.skaffold/config
global:
  local-cluster: false
  survey:
    last-prompted: "2021-03-20T13:42:49+05:30"
  collect-metrics: true
```

Similarly, you can unset the configuration using the following commands:

```
$ skaffold config unset --global local-cluster
unset global value local-cluster
$ cat ~/.skaffold/config
global:
  survey:
    last-prompted: "2021-03-20T13:42:49+05:30"
  collect-metrics: true
kubeContexts: []
```

In this section, we covered the components of the skaffold.yaml configuration file. We also looked at some of the global configuration settings that you can set via the Skaffold CLI commands.

Summary

This chapter introduced you to some of Skaffold's peculiarities, such as super-fast local development, effortless remote development, built-in tag management, lightweight capability, and file sync capability to name a few. These are compelling features that will help you to improve the developer experience. Additionally, we looked at the Skaffold architecture and discovered that Skaffold has a pluggable architecture. This means that you can always bring your own tools to build and deploy your applications. Following this, we covered the steps that typically occur within the Skaffold development workflow. Finally, at the end of the chapter, we studied the Skaffold main components and some global configurations supported via the Skaffold configuration.

In this chapter, the main goal was to give you an insight into Skaffold's features and internals by looking at its architecture and typical development workflow. You have developed a deep understanding of Skaffold, and now it will be easier for you to connect the dots between the previous and upcoming chapters.

In the next chapter, we will go through the different ways of installing Skaffold. Additionally, we will explore the Skaffold CLI commands.

References

- The official Skaffold documentation (`https://skaffold.dev/docs/`)

5

Installing Skaffold and Demystifying Its Pipeline Stages

In the previous chapter, we took a deep dive into Skaffold's architecture and workflow. We also looked at Skaffold's configuration. This chapter will cover how to install Skaffold on different operating systems, such as Linux, Windows, and macOS. We will also explore common CLI commands and how to use these commands in Skaffold's different pipeline stages.

In this chapter, we're going to cover the following main topics:

- Installing Skaffold

- Understanding common CLI commands

- Understanding Skaffold's pipeline stages

- Debugging with Skaffold

By the end of this chapter, you will know how to install Skaffold on different platforms. You will also gain a solid understanding of the most used CLI commands for Skaffold.

Technical requirements

To follow along with the examples of this chapter, you will need the following:

- The Skaffold CLI (https://skaffold.dev/docs/install/)
- minikube (https://minikube.sigs.k8s.io/docs/) or Docker Desktop for macOS and Windows (https://www.docker.com/products/dockerdesktop)

Installing Skaffold

Skaffold, being a CLI tool, needs to be downloaded and installed first on your favorite operating system. The following are the supported platforms where you can download and install Skaffold:

- Linux
- macOS
- Windows
- Docker
- Google Cloud SDK

Let's discuss these options in detail.

Installing Skaffold on Linux

For Linux, you can use the following URLs to download the latest stable release of Skaffold:

- https://storage.googleapis.com/skaffold/releases/latest/skaffold-linux-amd64
- https://storage.googleapis.com/skaffold/releases/latest/skaffold-linux-arm64

After downloading the binary, you can add it to your PATH variable. Alternatively, you can use the following commands.

For Linux on AMD64, use the following command:

```
curl -Lo skaffold https://storage.googleapis.com/skaffold/
releases/latest/skaffold-linux-amd64 && \sudo install skaffold
/usr/local/bin/
```

For Linux on ARM64, use the following command:

```
curl -Lo skaffold https://storage.googleapis.com/skaffold/
releases/latest/skaffold-linux-arm64 && \sudo install skaffold
/usr/local/bin/
```

There is also a bleeding-edge version of Skaffold, which is built with the latest commit. It may not be a stable version, so be careful while working with it. You can use the following URLs to download the bleeding edge version of Skaffold.

For Linux on AMD64, do the following:

```
curl -Lo skaffold https://storage.googleapis.com/skaffold/
builds/latest/skaffold-linux-amd64 && \sudo install skaffold /
usr/local/bin/
```

For Linux on ARM64, do the following:

```
curl -Lo skaffold https://storage.googleapis.com/skaffold/
builds/latest/skaffold-linux-arm64 && \sudo install skaffold /
usr/local/bin/
```

In this section, we looked at the commands for installing Skaffold on the Linux **operating system (OS)**.

Installing Skaffold on macOS

For macOS, you can use the following URLs to download the latest stable release of Skaffold:

- `https://storage.googleapis.com/skaffold/releases/latest/skaffold-darwin-amd64`

- `https://storage.googleapis.com/skaffold/releases/latest/skaffold-darwin-arm64`

After downloading the binary, you can add it to your `PATH` variable. Alternatively, you can use the following commands.

For macOS on AMD64, use the following command:

```
curl -Lo skaffold https://storage.googleapis.com/skaffold/
releases/latest/skaffold-darwin-amd64 && \sudo install skaffold
/usr/local/bin/
```

For macOS on ARM64, use the following command:

```
curl -Lo skaffold https://storage.googleapis.com/skaffold/
releases/latest/skaffold-darwin-amd64 && \sudo install skaffold
/usr/local/bin/
```

To download the build with the latest commit, you can use the following commands.

For macOS on AMD64, use the following command:

```
curl -Lo skaffold https://storage.googleapis.com/skaffold/
builds/latest/skaffold-darwin-amd64 && \sudo install skaffold /
usr/local/bin/
```

For macOS on ARM64, use the following command:

```
curl -Lo skaffold https://storage.googleapis.com/skaffold/
builds/latest/skaffold-darwin-amd64 && \sudo install skaffold /
usr/local/bin/
```

For macOS specifically, you can download Skaffold using the following package managers.

For Homebrew, use the following command:

```
brew install skaffold
```

For MacPorts, use the following command:

```
sudo port install skaffold
```

In this section, we explored various commands for installing Skaffold on macOS.

Installing Skaffold on Windows

For Windows, you can use the following URL to download the latest stable release of Skaffold:

```
https://storage.googleapis.com/skaffold/releases/latest/
skaffold-windows-amd64.exe
```

After downloading the EXE file, you can add it to your PATH variable.

To download the build with the latest commit, you can use the following URL:

```
https://storage.googleapis.com/skaffold/builds/latest/
skaffold-windows-amd64.exe
```

For Windows specifically, you can download Skaffold using the following Chocolatey package manager command:

```
choco install -y skaffold
```

The following is the output:

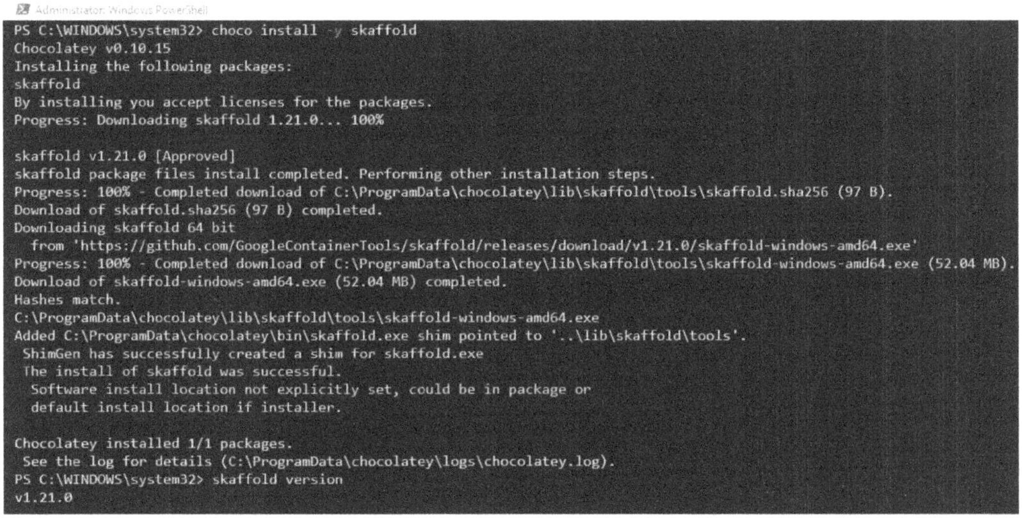

```
Administrator. Windows PowerShell
PS C:\WINDOWS\system32> choco install -y skaffold
Chocolatey v0.10.15
Installing the following packages:
skaffold
By installing you accept licenses for the packages.
Progress: Downloading skaffold 1.21.0... 100%

skaffold v1.21.0 [Approved]
skaffold package files install completed. Performing other installation steps.
Progress: 100% - Completed download of C:\ProgramData\chocolatey\lib\skaffold\tools\skaffold.sha256 (97 B).
Download of skaffold.sha256 (97 B) completed.
Downloading skaffold 64 bit
  from 'https://github.com/GoogleContainerTools/skaffold/releases/download/v1.21.0/skaffold-windows-amd64.exe'
Progress: 100% - Completed download of C:\ProgramData\chocolatey\lib\skaffold\tools\skaffold-windows-amd64.exe (52.04 MB).
Download of skaffold-windows-amd64.exe (52.04 MB) completed.
Hashes match.
C:\ProgramData\chocolatey\lib\skaffold\tools\skaffold-windows-amd64.exe
Added C:\ProgramData\chocolatey\bin\skaffold.exe shim pointed to '..\lib\skaffold\tools'.
 ShimGen has successfully created a shim for skaffold.exe
The install of skaffold was successful.
  Software install location not explicitly set, could be in package or
  default install location if installer.

Chocolatey installed 1/1 packages.
 See the log for details (C:\ProgramData\chocolatey\logs\chocolatey.log).
PS C:\WINDOWS\system32> skaffold version
v1.21.0
```

Figure 5.1 – Installing Skaffold on Windows

> **Note**
>
> There is a known issue (`https://github.com/chocolatey/shimgen/issues/32`) with the `skaffold dev` command where Skaffold doesn't clean up deployments when you press *Ctrl + C* on Windows when installed with Chocolatey package manager. The problem is not related to Skaffold but how Chocolatey interferes with *Ctrl + C* handling during installation.

This section covered how to install Skaffold on Windows.

Installing Skaffold using Docker

You can also download and run Skaffold inside a Docker container. To do that, you can use the following `docker run` command:

```
docker run gcr.io/k8s-skaffold/skaffold:latest skaffold
<command>
```

To work with an edge build with the latest commit, you can use the following command:

```
docker run gcr.io/k8s-skaffold/skaffold:edge skaffold <command>
```

I want to highlight one point regarding using a docker image for Skaffold. The Docker image's size is around ~3.83 GB, and this seems ridiculously large for Skaffold since in *Chapter 3, Skaffold – Easy-Peasy Cloud-Native Kubernetes Application Development*, we learned that Skaffold's binary size is around ~63 MB. This can be seen in the following screenshot:

```
ashish@MacBook-Air Chapter 3 % docker images gcr.io/k8s-skaffold/skaffold
REPOSITORY                      TAG       IMAGE ID       CREATED       SIZE
gcr.io/k8s-skaffold/skaffold    latest    2456bbf87f9a   3 weeks ago   3.83GB
```

Figure 5.2 – Skaffold Docker image size

So, why is the image size so large? This is because the image contains other tools as well, such as the gcloud SDK, kind, minikube, k3d, kompose, and bazel, to name a few.

You can verify what's inside your container image using the Dive CLI.

> **Tip**
> Dive allows you to check your image layer's contents and suggest different ways to shrink your image's size if you are wasting any space.

You can download Dive by following the instructions at `https://github.com/wagoodman/dive#installation`. Run the following command to get an inside view of your container image:

```
$ dive image tag/id/digest
```

The following is the output for the Skaffold docker image, which contains an image layer:

```
┃ ● Layers ┠──────────────────────────────────────────────────────────────
Cmp   Size  Command
      63 MB  FROM 4174d2bf0b4faba
      17 MB  bazel build ...
       0 B   /tmp/pkginstall/installer.sh
     190 MB  RUN /bin/sh -c apt-get update &&      apt-get install --no-install-recommends
      61 MB  COPY /usr/local/bin/docker /usr/local/bin/ # buildkit
      44 MB  COPY kubectl /usr/local/bin/ # buildkit
      42 MB  COPY helm /usr/local/bin/ # buildkit
      35 MB  COPY kustomize /usr/local/bin/ # buildkit
      25 MB  COPY kompose /usr/local/bin/ # buildkit
      15 MB  COPY container-structure-test /usr/local/bin/ # buildkit
      47 MB  COPY bazel /usr/local/bin/ # buildkit
     461 MB  COPY google-cloud-sdk/ /google-cloud-sdk/ # buildkit
     7.4 MB  COPY kind /usr/local/bin/ # buildkit
      22 MB  COPY k3d /usr/local/bin/ # buildkit
     152 MB  RUN /bin/sh -c bazel version # buildkit
     6.5 MB  RUN /bin/sh -c /google-cloud-sdk/install.sh     --usage-reporting=false
      25 kB  RUN /bin/sh -c gcloud auth configure-docker # buildkit
     1.1 GB  RUN /bin/sh -c apt-get update && apt-get install --no-install-recommends --n
     364 MB  COPY /usr/local/go /usr/local/go # buildkit
       0 B   #(nop) WORKDIR /skaffold
      72 MB  #(nop) COPY dir:1504f7130a654c3e7f30c2e813f8a056296ef1da45a2952469d0aef71393
     1.1 GB  |1 VERSION=v1.21.0 /bin/sh -c make clean out/skaffold VERSION=$VERSION && mv
       0 B   |1 VERSION=v1.21.0 /bin/sh -c rm -rf secrets $SECRET cmd/skaffold/app/cmd/st
     994 kB  |1 VERSION=v1.21.0 /bin/sh -c skaffold credits -d /THIRD_PARTY_NOTICES
```

Figure 5.3 – Skaffold Docker image layers

As you can see from the layers inside the image, we have many tools available, not just Skaffold. Another advantage of using this Docker image is that you don't have to install these tools separately, and you can use the same image to play or experiment with these tools.

This section covered how to install Skaffold using a Docker image.

Installing Skaffold using gcloud

Google developed Skaffold, so it nicely fits into the Google products ecosystem. If you have **Google's Cloud SDK** installed on your machine, you can use the `gcloud components install skaffold` command to install Skaffold.

We will cover how to install the gcloud SDK in *Chapter 8, Deploying a Spring Boot Application to Google Kubernetes Engine Using Skaffold*. For now, we can assume that the Cloud SDK is already installed. You can view the installed and not installed components using the `gcloud list` command. The following is the output:

```
ashish@MacBook-Air Chapter 3 % gcloud components list

Your current Cloud SDK version is: 331.0.0
The latest available version is: 335.0.0

┌─────────────────────────────────────────────────────────────────────────────────────────────┐
│                                          Components                                           │
├──────────────────┬──────────────────────────────────────────────────────┬──────────────────────────┬──────────┤
│      Status       │                         Name                          │            ID            │   Size   │
├──────────────────┼──────────────────────────────────────────────────────┼──────────────────────────┼──────────┤
│ Update Available │ BigQuery Command Line Tool                           │ bq                       │ < 1 MiB  │
│ Update Available │ Cloud SDK Core Libraries                             │ core                     │ 17.8 MiB │
│ Update Available │ Cloud Storage Command Line Tool                      │ gsutil                   │ 3.9 MiB  │
│ Not Installed    │ App Engine Go Extensions                             │ app-engine-go            │ 4.8 MiB  │
│ Not Installed    │ Appctl                                               │ appctl                   │ 18.5 MiB │
│ Not Installed    │ Cloud Bigtable Command Line Tool                     │ cbt                      │ 7.6 MiB  │
│ Not Installed    │ Cloud Bigtable Emulator                              │ bigtable                 │ 6.6 MiB  │
│ Not Installed    │ Cloud Datalab Command Line Tool                      │ datalab                  │ < 1 MiB  │
│ Not Installed    │ Cloud Datastore Emulator                             │ cloud-datastore-emulator │ 18.4 MiB │
│ Not Installed    │ Cloud Firestore Emulator                             │ cloud-firestore-emulator │ 41.9 MiB │
│ Not Installed    │ Cloud Pub/Sub Emulator                               │ pubsub-emulator          │ 60.4 MiB │
│ Not Installed    │ Cloud SQL Proxy                                      │ cloud_sql_proxy          │ 7.4 MiB  │
│ Not Installed    │ Emulator Reverse Proxy                               │ emulator-reverse-proxy   │ 14.5 MiB │
│ Not Installed    │ Google Cloud Build Local Builder                     │ cloud-build-local        │ 6.2 MiB  │
│ Not Installed    │ Google Container Registry's Docker credential helper │ docker-credential-gcr    │ 2.2 MiB  │
│ Not Installed    │ Kustomize                                            │ kustomize                │ 22.8 MiB │
│ Not Installed    │ Minikube                                             │ minikube                 │ 23.7 MiB │
│ Not Installed    │ Nomos CLI                                            │ nomos                    │ 22.5 MiB │
│ Not Installed    │ On-Demand Scanning API extraction helper             │ local-extract            │ 11.5 MiB │
│ Not Installed    │ Skaffold                                             │ skaffold                 │ 17.5 MiB │
└──────────────────┴──────────────────────────────────────────────────────┴──────────────────────────┴──────────┘
```

Figure 5.4 – gcloud list command output

From the preceding output, it is clear that Skaffold is not installed. It is not mandatory but before we move on to the installation, make sure that `gcloud` has been installed and that its components are up to date. We can run the following command to do this:

```
gcloud components update
```

Finally, we can install Skaffold using the following `gcloud` command:

```
gcloud components install skaffold
```

The following is the output:

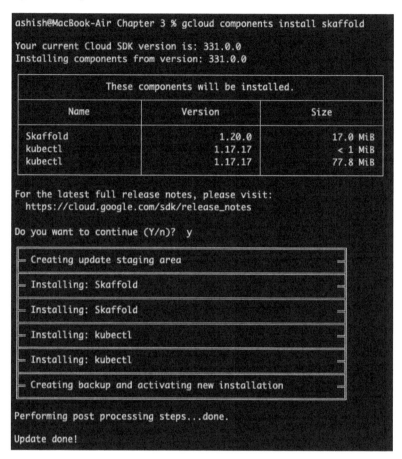

Figure 5.5 – Installing Skaffold via gcloud

In this section, we discussed different ways to install Skaffold. Now, let's discuss the Skaffold CLI commands.

Understanding common CLI commands

So far, we have introduced you to commands such as `skaffold dev` and `skaffold run`, but there are many such commands that you can use to either create an end-to-end pipeline or use individually in your CI/CD pipelines. We will categorize these commands into the following sections. You can also discover the supported options for these commands by enabling the `skaffold completion bash/zsh` command and pressing *Tab* after entering the command:

- **Commands for end-to-end pipelines:**

  - `skaffold run`: This command allows you to build and deploy once.

  - `skaffold dev`: This command allows you to trigger the continuous development loop for build and deploy. This workflow will clean up on exit.

  - `skaffold debug`: This command allows you to trigger the continuous development loop for build and deploy pipelines *in debug mode*. This workflow will also clean up on exit.

- **Commands for CI/CD pipelines:**

  - `skaffold build`: This command allows you to just build, tag, and push your image(s).

  - `skaffold test`: This command allows you to run tests against your built application images.

  - `skaffold deploy`: This command allows you to deploy the given image(s).

  - `skaffold delete`: This command allows you to clean up the deployed artifacts.

  - `skaffold render`: This command allows you to build your application images and then export the hydrated (with a newly built image tag) Kubernetes manifests to a file or terminal.

  - `skaffold apply`: This command takes templated Kubernetes manifests as input and creates resources on the target cluster.

- **Commands for getting started:**

  - `skaffod init`: This command allows you to bootstrap the Skaffold config.

  - `skaffold fix`: This command allows you to upgrade the schema version.

- **Miscellaneous commands**:

  - `skaffold help`: This command allows you to print help. Use `skaffold options` to get a list of global command-line options (this applies to all commands).

  - `skaffold version`: This command allows you to get Skaffold's version.

  - `skaffold completion`: This command allows you to set up tab completion for the CLI. It supports the same options as `skaffold version`.

  - `skaffold config`: This command allows you to manage context-specific parameters. It supports the same options as `skaffold version`.

  - `skaffold credits`: This command allows you to export third-party notices to a given path (`./skaffold-credits` by default). It supports the same options as `skaffold version`.

  - `skaffold diagnose`: This command allows you to run diagnostics of Skaffold works in your project.

  - `skaffold schema`: This command allows you to list and print JSON schemas used to validate `skaffold.yaml` configuration. It supports the same options as `skaffold version`.

In this section, we talked about Skaffold commands and their usage. In the next section, we will try to understand Skaffold's different pipeline stages.

Understanding Skaffold pipeline stages

So far, we have built a basic understanding of how Skaffold works. From the previous chapters, we know that it picks source code changes in your project and creates container images with the tool of your choice; the images, once successfully built, are tagged as you see fit and pushed to the repository you specify. Skaffold also helps you deploy the artifacts to your Kubernetes cluster at the end of the workflow, once again using the tools you prefer.

Skaffold permits you to skip stages. If, for example, you run Kubernetes locally with Minikube or Docker desktop, Skaffold is smart enough to make that choice for you and will not push artifacts to a remote repository.

Let's look at Skaffold's pipeline stages in detail to understand what other choices we have in each pipeline stage. Skaffold pipeline stages can be broadly classified into the following areas:

- Init
- Build
- Tag
- Test
- Deploy
- File
- Log tailing
- Port forwarding
- Cleanup

Let's discuss each in detail.

Init stage

In this stage, we typically generate a basic Skaffold configuration file to get your project up and running in seconds. Skaffold looks at your project directory for any build configuration files, such as `Dockerfile`, `build.gradle`, and `pom.xml`, and then auto-generates a build and deploy config.

Skaffold currently supports build detection for the following builders:

- Docker
- Jib
- Buildpacks

If Skaffold detects multiple build configuration files, it will prompt you to pair your build configuration files with any images that have been detected in your deploy configuration.

> **Tip**
> Starting with Skaffold v1.27.0, you no longer have to provide the `XXenableJibInit` or `XXenableBuildpacksInit` flag with the `skaffold init` command as their default values are set to `true`. This also means that the `init` command will detect if you should use Jib or Buildpacks based upon your project, without the need to specify these flags.

For example, you might be asked to choose from the following after you run the `skaffold init` command. In this example, we have a `Dockerfile` in the root directory, so Skaffold is asking you to choose the build configuration for this project:

```
ashish@MacBook-Air chapter02 % skaffold init
? Choose the builder to build image docker.io/hiashish/helloworld  [Use arrows to move, type to filter]
  Buildpacks (pom.xml)
> Docker (Dockerfile)
  None (image not built from these sources)
```

Figure 5.6 – skaffold init prompt

Similarly, for deployment, Skaffold will look at your project directory and if it detects some Kubernetes manifests – that is, `deployment.yaml` or `sevice.yaml` – is already present, then it will automatically add those to the `deploy` section of the `skaffold.yaml` file:

```
? Choose the builder to build image docker.io/hiashish/helloworld Docker (Dockerfile)
? Which builders would you like to create kubernetes resources for?
apiVersion: skaffold/v2beta18
kind: Config
metadata:
  name: chapter-
build:
  artifacts:
  - image: docker.io/hiashish/helloworld
    docker:
      dockerfile: Dockerfile
deploy:
  kubectl:
    manifests:
    - k8s/mydeployment.yaml
    - k8s/myservice.yaml

? Do you want to write this configuration to skaffold.yaml? (y/N) Y
```

Figure 5.7 – Generating a Skaffold configuration file

If you don't have the manifests ready but want Skaffold to handle the manifest generation part, then don't worry – you can pass the `--generate-manifests` flag with the `skaffold init` command.

Build stage

Skaffold supports various tools for image building.

From the following table, you can understand that image building can be done locally, in a cluster, or using Google Cloud Build remotely:

Container Image Builders	Local Build	In Cluster Build	Remote on Google Cloud Build
Dockerfile	Yes	Yes	Yes
Jib Maven and Gradle	Yes	NA	Yes
Cloud Native Buildpacks	Yes	NA	Yes
Bazel	Yes	NA	NA
Custom Script	Yes	Yes	NA

Table 5.1– Skaffold supported container image builders

We will learn more about these options in *Chapter 6, Working with Skaffold Container Image Builders and Deployers*. In a cluster, builds are supported by kaniko or using a custom script. Remote builds are only supported for Dockerfile, Jib, and Buildpacks using Cloud Build. For a local build, you can pretty much use any of the supported image building methods.

You can set a build configuration through the `build` section of the `skaffold.yaml` file. The following is an example of this:

```
build:
  artifacts:
    - image: docker.io/hiashish/skaffold-introduction
      jib: {}
```

Now that we have covered the build stage, next, we will take a look at the tag stage.

Tag stage

Skaffold supports the following image tagging strategies:

- Tagging is available through the `gitCommit tagger`, which utilizes Git commits to tag images.

- Tagging is available through the `sha256 tagger`, which uses the latest tag to tag images.

- Tagging is available through the `envTemplate tagger`, which uses **Environment Variables** to tag images.

- Tagging is available through the `dateTime tagger`, which accepts the current **Date and Time** with a configurable pattern.

- Tagging is available through the `customTemplate tagger`, which uses a combination of the existing taggers as components in a template.

An image tagging strategy can be configured using the `tagPolicy` field in the `build` section of `skaffold.yaml`. If no `tagPolicy` is specified, then the default is the `gitCommit` strategy. Please refer to the following code snippet:

```
build:
  artifacts:
    - image: docker.io/hiashish/skaffold-introduction
      jib: {}
  tagPolicy:
    sha256: {}
```

Now that we know about the different image tagging strategies with Skaffold, we will go through the test stage.

Test stage

Skaffold has an integration testing phase between build and deploy. It supports container structure tests and custom tests for integration testing. Let's discuss them in detail.

Container structure test

Skaffold provides support for running container structure tests (`https://github.com/GoogleContainerTools/container-structure-test`) on container images we build with Skaffold. The Container Structure Test framework primarily aims to verify the contents and structure of the container. For example, we may want to run some command inside a container to test whether it gets executed successfully or not. We can define tests per image in the Skaffold config. After building the artifact, Skaffold will run the associated structure tests on that image. If the tests fail, Skaffold will not proceed with the deployment.

Custom test

With a Skaffold custom test, developers can run custom commands as part of their development loop. The custom test will run before deploying the image to the Kubernetes cluster. The command will execute on the local machine where Skaffold is being executed and works with all supported Skaffold platforms. You can opt out of running custom tests by using the `--skip-tests` flag. You can run tests individually with the `skaffold test` command.

The following are some of the use cases for a custom test:

- Running unit tests

- Running validation and security scans on images using GCP Container Analysis or Anchore Grype

- We can also validate Kubernetes manifests before deployment using tools like **kubeval** (`https://github.com/instrumenta/kubeval`) or **kubeconform** (`https://github.com/yannh/kubeconform`).

- In the case of Helm charts, we can use the **helm lint** command before deployment.

In the following example, we have a profile named `test` and we are using the `mvn test` command to run various tests. We will be using the `skaffold dev --profile=test` command here, which runs tests after the build and before deployment:

```
profiles:
  - name: test
    test:
      - image: docker.io/hiashish/skaffold-introduction
        custom:
          - command: mvn test -Dmaven.test.skip=false
```

In the logs, you will see the following, which states that the tests have been started and that there are no failures:

```
Starting test...
Testing images...
Running custom test command: "mvn test -Dmaven.test.skip
=false"
[INFO] Results:
[INFO]
[INFO] Tests run: 5, Failures: 0, Errors: 0, Skipped: 0
```

With that, we have learned how we can execute custom tests with Skaffold. In the deploy stage, we will learn about deploying an application with Skaffold.

Deploy stage

Skaffold deploy stage typically renders the Kubernetes manifests by replacing the untagged image names in the Kubernetes manifests with the final tagged image names. It also might go through the additional intermediary step of expanding templates for helm or calculating overlays for kustomize. Then, Skaffold will deploy the final Kubernetes manifests to the cluster. And to make sure the deployment happens, ideally, it will wait for the deployed resources to stabilize by doing health checks.

Health checks are enabled by default and are a great feature for your CI/CD pipeline use cases to make sure that deployed resources are healthy and we can proceed further in the pipeline. Skaffold internally uses the `kubectl rollout status` command to test the status of the deployment.

For example, in the following logs, you can see that Skaffold waited for the deployment to stabilize:

```
Starting test...
Tags used in deployment:
 - docker.io/hiashish/skaffold-introduction -> docker.io/
hiashish/skaffold-introduction:fcda757-dirty@sha256:f07c1dc192
cf5f391a1c5af8dd994b51f7b6e353a087cbcc49e754367c8825cc
Starting deploy...
 - deployment.apps/skaffold-introduction created
 - service/skaffold-introduction created
Waiting for deployments to stabilize...
 - deployment/skaffold-introduction: 0/4 nodes are available: 2
Insufficient memory, 4 Insufficient cpu.
    - pod/skaffold-introduction-59b479ddcb-f8ljj: 0/4 nodes are
available: 2 Insufficient memory, 4 Insufficient cpu.
 - deployment/skaffold-introduction is ready.
Deployments stabilized in 56.784 seconds
Press Ctrl+C to exit
Watching for changes...
```

Skaffold currently supports the following tools for deploying applications to local or remote Kubernetes clusters:

- `kubectl`
- `helm`
- `kustomize`

You can set the deploy configuration through the `deploy` section of the `skaffold.yaml` file, as shown here:

```
deploy:
  kubectl:
    manifests:
      - k8s/mydeployment.yaml
      - k8s/myservice.yaml
```

With that, we have learned how we can deploy an image to Kubernetes using Skaffold. Next, we will explore how to sync changes directly to a pod without rebuilding and redeploying an image using file sync.

File sync

Skaffold has a great feature through which it can copy the changed files to a deployed container, without the need to rebuild, redeploy, and restart the corresponding pod. We can enable this file copying feature by adding a `sync` section with sync rules to the artifact in the `skaffold.yaml` file. Internally, Skaffold creates a `.tar` file with changed files that match the sync rules we define in the `skaffold.yaml` file. Then, this `.tar` file is transferred to and extracted inside the corresponding containers.

Skaffold supports the following types of sync:

- `manual`: In this mode, we need to specify the source file path from the local and destination path of the running container.
- `infer`: In this mode, Skaffold will infer the destination path by looking at your Dockerfile. Under sync rules, you can specify which files are eligible for syncing.
- `auto`: In this mode, Skaffold will automatically generate sync rules for the known file types.

To understand the **file sync** functionality, we will use the Spring Boot application that we built in *Chapter 3, Skaffold – Easy-Peasy Cloud-Native Kubernetes Application Development*. The Spring Boot application exposes a `/states` REST endpoint that will return all Indian states and their capitals. We have added a new profile named sync to the `skaffold.yaml` file.

In the following `skaffold.yaml` file, we have used `jib` as an image builder. Jib integration with Skaffold allows you to auto-sync your class files, resource files, and Jib's extra directories files to a remote container once changes have been made. However, it can only be used with Jib in the default build mode (exploded) for non-WAR applications because of some limitations. You also need to have the Spring Boot Developer Tools dependency in your project for this to work. It will also work with any embedded server that can do a reload or restart:

```yaml
apiVersion: skaffold/v2beta20
kind: Config
metadata:
  name: skaffold-introduction
build:
  artifacts:
    - image: docker.io/hiashish/skaffold-introduction
      jib: { }
deploy:
  kubectl:
    manifests:
      - k8s/mydeployment.yaml
      - k8s/myservice.yaml
profiles:
  - name: sync
    build:
      artifacts:
        - image: docker.io/hiashish/skaffold-introduction
          jib: {}
          sync:
            auto: true
```

In the Spring Boot application, we deliberately made a mistake by changing the name of Bengaluru to Bangalore. In the output, you will see the following after running the `skaffold dev --profile=sync` command:

```
{"name":"Karnataka","capital":"Bangalore"}
```

Figure 5.8 – Output before sync

Now, since we have Jib's auto-sync set to `true`, any changes that are made to the `schema.sql` file will be directly synced with the pod running inside the Kubernetes cluster. We made changes in the `schema.sql` file, and they were synced directly with a running pod by just restarting the application. Here, we don't have to rebuild the image, push the image, redeploy the image, or restart the pod. You will see the following output in the streamed logs on your console after you make this change:

```
: Completed initialization in 3 ms
[skaffold-introduction] 2021-07-18 21:07:03.279  INFO 1 ---
[nio-8080-exec-1] c.p.c.indianstates.StateController        :
Getting all states.
Syncing 1 files for docker.io/hiashish/skaffold-
introduction:fcda757-dirty@sha256:f07c1dc192cf5f391a1c5af8d
d994b51f7b6e353a087cbcc49e754367c8825cc
Watching for changes...
```

After hitting the URL again, you will see the changed output:

```
{"name":"Karnataka","capital":"Bengaluru"}
```

Figure 5.9 – Output after sync

`schema.sql` was under our resources, so let's see whether, when we make changes to a Java class file, those will also be synced. Let's try it out.

To test this, I will tweak the logging statement we have in the `StateController` class. We have the following log statement:

```
LOGGER.info("Getting all states.");
```

We will change it to the following:

```
LOGGER.info("Getting all Indian states and their capitals.");
```

After making these changes, you should see the following in the streamed logs on your console. You might be wondering why five files have been synced since we only changed one file. Well, the reason for this is that Jib transferred the entire layer, which contains your class file:

```
: Completed initialization in 3 ms
[skaffold-introduction] 2021-07-18 21:19:52.941  INFO 1 ---
[nio-8080-exec-2] c.p.c.indianstates.StateController        :
Getting all states.
```

```
Syncing 5 files for docker.io/hiashish/skaffold-
introduction:fcda757-dirty@sha256:f07c1dc192cf5f391a1c5af

8dd994b51f7b6e353a087cbcc49e754367c8825cc

Watching for changes...
```

Similarly, in the streamed logs, we will see the changed logging statement:

```
[skaffold-introduction] 2021-07-18 21:40:46.868  INFO 1 ---
[nio-8080-exec-1] c.p.c.indianstates.StateController       :
Getting all Indian states and their capitals.
```

With that, we have learned about the direct file sync capabilities of Skaffold. Now, let's understand how we can tail logs with various Skaffold commands.

Log tailing

Skaffold can tail logs for containers that have been built and deployed by it. With this feature, you can tail logs from your cluster to your local machine when you execute `skaffold dev`, `skaffold debug`, or `skaffold run`.

Log tailing is enabled for the `skaffold dev` and `skaffold debug` modes by default. For skaffold `run`, you can use the `-tail` flag to explicitly enable log tailing.

For a typical Spring Boot application, you will see the following in the tail logs after you've built and deployed it using `skaffold dev`.

In the following log, you can see that after successfully building and deploying to the cluster, the application logs are streamed to the console:

```
Starting test...
Tags used in deployment:
 - docker.io/hiashish/skaffold-introduction -> docker.io/
hiashish/skaffold-introduction:fcda757-dirty@sha256:f07c1dc1
92cf5f391a1c5af8dd994b51f7b6e353a087cbcc49e754367c8825cc
Starting deploy...
 - deployment.apps/skaffold-introduction created
 - service/skaffold-introduction created
Waiting for deployments to stabilize...
 - deployment/skaffold-introduction: 0/4 nodes are available: 2
Insufficient memory, 4 Insufficient cpu.
    - pod/skaffold-introduction-59b479ddcb-f8ljj: 0/4 nodes are
available: 2 Insufficient memory, 4 Insufficient cpu.
```

```
 - deployment/skaffold-introduction is ready.
Deployments stabilized in 56.784 seconds
Press Ctrl+C to exit
Watching for changes...
[skaffold-introduction]
[skaffold-introduction] 2021-07-18 21:06:44.072  INFO 1 ---
[ restartedMain] c.p.c.i.IndianStatesApplication        :
Starting IndianStatesApplication using Java 16-ea on skaffold-
introduction-59b479ddcb-f8ljj with PID 1 (/app/classes started
by root in /)
```

At this point, we know how we can tail logs from a running container with Skaffold. Next, let's discuss port forwarding with Skaffold.

Port forwarding

Skaffold supports automatic port forwarding of services and user-defined port forwards in dev, debug, deploy, or run mode. You don't have to expose an endpoint to access your application. Port forwarding is helpful for local development. Skaffold uses `kubectl port-forward` internally to implement port forwarding. You can define your custom port forward explicitly in `skaffold.yaml` or pass the `--port-forward` flag while running `skaffold dev`, `debug`, `run`, or `deploy`.

The following is an example of user-defined port forwarding. In this example, Skaffold will try to forward port `8080` to `localhost:9000`. If port `9000` is unavailable for some reason, then Skaffold will forward to a random open port:

```
profiles:
  - name: userDefinedPortForward
    portForward:
      - localPort: 9090
        port: 8080
        resourceName: reactive-web-app
        resourceType: deployment
```

It is good practice to clean up the resources we create using Skaffold after our work is completed. Now, let's learn how we can clean up and delete Kubernetes resources with Skaffold.

Cleanup

With the skaffold run and skaffold dev commands, we can create resources in the Kubernetes cluster, create images stored on the local Docker daemon, and sometimes push images to remote registry. Doing all this work can have side effects on your local and deployment environments, in that you might fill up a significant amount of disk space in your local environment.

Skaffold provides a cleanup functionality to neutralize some of these side effects:

- You can clean up Kubernetes resources by running skaffold delete, or perform an automatic cleanup by using *Ctrl + C* for skaffold dev and skaffold debug.

- You can enable image pruning for local Docker daemon images by passing the --no-prune=false flag. Since artifact caching is enabled by default, you need to disable that for the purge to work. The actual command you need to run is skaffold dev --no-prune=false --cache-artifacts=false. By pressing *Ctrl + C* for skaffold dev and skaffold debug, Skaffold will automatically clean the images stored on the local Docker daemon.

- For images that have been pushed to remote container registries, the user has to take care of the cleanup.

For example, to test image pruning, we can use the following docker profile to build images using our local Docker daemon:

```
- name: docker
  build:
    artifacts:
      - image: docker.io/hiashish/skaffold-introduction
        docker:
          dockerfile: Dockerfile
```

Then, we can run the skaffold dev --no-prune=false --cache-artifacts=false command. After the build and deployment, we can press *Ctrl + C*, which should prune the images and delete any Kubernetes resources as well. In the following logs, you can see that after pressing *Ctrl + C*, Skaffold started deleting Kubernetes resources and pruned images:

```
Cleaning up...
 - deployment.apps "skaffold-introduction" deleted
 - service "skaffold-introduction" deleted
Pruning images...
```

In this section, we took a deep dive into Skaffold pipeline stages such as init, build, and deploy, to name a few. In the next section, we will talk about debugging an application that's been deployed to a Kubernetes cluster with Skaffold.

Debugging with Skaffold

Skaffold supports debugging containerized applications running on Kubernetes with the `skaffold debug` command. Skaffold provides debugging for different container's runtime technology. Once debugging has been enabled, the associated debugging ports are exposed and labeled to be port-forwarded to the local machine. IntelliJ IDE's plugins, like Cloud Code, internally use Skaffold to add and attach the correct debugger for your language.

However, in debug mode, `skaffold debug` will disable image rebuilding and syncing as it can lead to debugging sessions terminating accidentally if we save file changes. You can allow image rebuilding and syncing with the `--auto-build`, `--auto-deploy`, and `--auto-sync` flags.

The `skaffold debug` command supports the following languages and runtimes:

- Go 1.13+ (runtime ID: go) and using Delve
- Node.js (runtime ID: nodejs) and using the Node.js Inspector Chrome DevTools
- The Java and JVM languages (runtime ID: jvm) and using JDWP
- Python 3.5+ (runtime ID: python) and using `debugpy` (Debug Adapter Protocol) or `pydevd`
- .NET Core (runtime ID: netcore) using `vsdbg`

In the IDE, like IntelliJ, you need to add the Remote Java Application configuration to your Run/Debug configurations once you start your application. You must also select the port/address you defined when starting your application. Then, you are ready to debug:

```
[skaffold-introduction] Picked up JAVA_TOOL_OPTIONS:
-agentlib:jdwp=transport=dt_socket,server=y,
address=5005,suspend=n,quiet=y
Port forwarding pod/skaffold-introduction-766df7f799-dmq4t in
namespace default, remote port 5005 -> 127.0.0.1:5005
```

In IntelliJ, you should see the following after setting up the breakpoint. On the breakpoint, you should see the tick mark once the debug session has been activated:

```
@GetMapping(◎✔"/states")
private List<State> getAllStates() {
    LOGGER.info("Getting all Indian states and their capitals.");
    return stateService.findAll();
}
```

Figure 5.10 – Breakpoint enabled

In the **Debug** console logs, you should see the following once the debug session has started. Now, you are ready to debug your application:

Figure 5.11 – Debugger attached

In this section, we took a deep dive into Skaffold's debugging capabilities. We also learned how to debug a containerized version of our application using the `skaffold debug` command. You can also debug using the Cloud Code IntelliJ plugin, which we will cover in *Chapter 7*, *Building and Deploying a Spring Boot Application with the Cloud Code Plugin*.

Summary

In this chapter, we started by discovering various ways to install Skaffold on different operation systems. We covered the installation for popular OSes such as macOS, Windows, and Linux. Then, we looked at various commands supported by Skaffold that help build and deploy the Kubernetes application. We also covered some miscellaneous and housekeeping commands. Then, we discovered the different Skaffold pipeline stages, such as init, build, and deploy to name a few. Finally, we discussed debugging an application with Skaffold with a command such as `skaffold dev`.

In the next chapter, we will discuss Skaffold container image builders (Dockerfile, kaniko, Buildpacks, Jib) and deployers (Helm, kubectl, kustomize).

Further reading

If you want to learn more about Skaffold, please take a look at its documentation at `https://skaffold.dev/docs/`.

6
Working with Skaffold Container Image Builders and Deployers

In the previous chapter, we took a deep dive into the Skaffold CLI and its pipeline stages. We also looked at the Skaffold configuration. In this chapter, we will introduce you to reactive programming by creating a Reactive Spring Boot CRUD application. We will then learn about Skaffold's pluggable architecture, which supports different methods of building and deploying container images to a Kubernetes cluster.

In this chapter, we're going to cover the following main topics:

- Creating a Reactive Spring Boot CRUD application
- Working with Skaffold container image builders
- Exploring Skaffold container image deployers

By the end of this chapter, you will have gained a solid understanding of Skaffold's supported container image builders, including Jib, Docker, and Buildpacks. You will also learn about Helm, kubectl, and Kustomize, which are supported by Skaffold to help deploy your containerized application to Kubernetes.

Technical requirements

To follow along with the examples in this chapter, you will need the following:

- Helm (`https://helm.sh/docs/intro/install/`)

- Kustomize (`https://kubectl.docs.kubernetes.io/installation/kustomize/`)

- Eclipse (`https://www.eclipse.org/downloads/`) or the IntelliJ IDE (`https://www.jetbrains.com/idea/download/`)

- Git (`https://git-scm.com/downloads`)

- Skaffold (`https://skaffold.dev/docs/install/`)

- Spring Boot 2.5

- OpenJDK 16

- minikube (`https://minikube.sigs.k8s.io/docs/`) or Docker Desktop for macOS and Windows (`https://www.docker.com/products/dockerdesktop`)

You can download the code examples for this chapter from this book's GitHub repository at `https://github.com/PacktPublishing/Effortless-Cloud-Native-App-Development-Using-Skaffold/tree/main/Chapter06`.

Creating a Reactive Spring Boot CRUD application

To demonstrate working with various container image builders that Skaffold supports, we will create a simple Reactive Spring Boot CRUD REST application. We expose a REST endpoint called `/employee` when the app is accessed locally through curl or a REST client such as Postman, which will return with employee data.

First, to build some context, let's discuss the reactive way of building an application. Reactive programming (`https://projectreactor.io/`) is a new way of building non-blocking applications that are asynchronous, event-driven, and require a small number of threads to scale. What also keeps them separate from typical non-reactive applications is that they can provide the backpressure mechanism to ensure producers don't overwhelm consumers.

Spring WebFlux is a reactive web framework that was introduced with Spring 5. Spring WebFlux doesn't require a servlet container and can be run on non-blocking containers such as Netty and Jetty. We would need to add the `spring-boot-starter-webflux` dependency to add support for Spring WebFlux. With Spring MVC, we have Tomcat as the default embedded server, while with WebFlux, we get Netty. Spring WebFlux controllers typically return reactive types; that is, Mono or Flux instead of collections or domain objects.

The following are the Maven dependencies that will be used for this Spring Boot application:

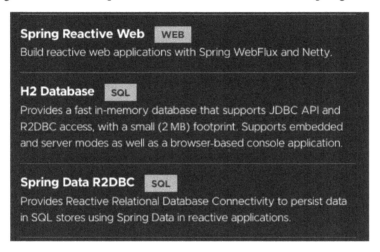

Figure 6.1 – Maven dependencies

Let's start by walking through the code of the application:

1. Here, we have an employee table with five columns: `id`, `first_name`, `last_name`, `age`, and `salary`. The `id` column is auto-incremented. The other columns follow the default snake case naming scheme. The following `schema.sql` SQL file is available from the `src/main/resources/schema.sql` path, in the source code directory:

    ```
    DROP TABLE IF EXISTS employee ;
    CREATE TABLE employee ( id SERIAL PRIMARY KEY, first_name
    VARCHAR(100) NOT NULL,last_name VARCHAR(100) NOT NULL,
    age integer,salary decimal);
    ```

    Since the H2 driver is on the classpath and we don't have to specify a connection URL, Spring Boot automatically starts an embedded H2 database at application startup.

2. To initialize the database schema at application startup, we also need to register `ConnectionFactoryInitializer` to pick up the `schema.sql` file, as mentioned in the following main class for our application. Here, we are also saving a few `Employee` entities that we can use later:

```
@SpringBootApplication
public class ReactiveApplication {
    private static final Logger logger =LoggerFactory.
      getLogger(ReactiveApplication.class);


    public static void main(String[] args) {
      SpringApplication.run(ReactiveApplication.class,
        args);
    }


    @Bean
    ConnectionFactoryInitializer initializer
      (ConnectionFactory connectionFactory) {
      ConnectionFactoryInitializer initializer = new
      ConnectionFactoryInitializer();
      initializer.setConnectionFactory
        (connectionFactory);
      initializer.setDatabasePopulator(new
      ResourceDatabasePopulator(new
      ClassPathResource("schema.sql")));
      return initializer;
    }


    @Bean
    CommandLineRunner init(EmployeeRepository
      employeeRepository) {
        return args -> {
            List<Employee> employees =  List.of(
                new Employee("Peter", "Parker", 25,
                    20000),
                new Employee("Tony", "Stark", 30,
                    40000),
```

```
                    new Employee("Clark", "Kent", 31,
                        60000),
                    new Employee("Clark", "Kent", 32,
                        80000),
                        new Employee("Bruce", "Wayne", 33,
                        100000)
                );
                logger.info("Saving employee " +
                  employeeRepository.saveAll
                    (employees).subscribe());
            };
        }
    }
```

3. With Spring Data R2DBC, you don't have to write an implementation of the repository interface as it creates an implementation for you at runtime. `EmployeeRepository` extends `ReactiveCrudRepository` and inherits various methods for saving, deleting, and finding employee entities using reactive types. Following is the CRUD repository:

```
import com.example.demo.model.Employee;
import org.springframework.data.repository.reactive.
Reactive
  CrudRepository;
    public interface EmployeeRepository extends
      ReactiveCrudRepository<Employee,Long> {
}
```

Following is the `EmployeeService` class:

```
import com.example.demo.model.Employee;
import com.example.demo.repository.EmployeeRepository;
import org.springframework.stereotype.Service;
import reactor.core.publisher.Flux;
import reactor.core.publisher.Mono;

@Service
public class EmployeeService {
```

```
    private final EmployeeRepository
      employeeRepository;

    public EmployeeService(EmployeeRepository
      employeeRepository) {
        this.employeeRepository = employeeRepository;
    }

    public Mono<Employee> createEmployee(Employee
      employee) {
        return employeeRepository.save(employee);
    }

    public Flux<Employee> getAllEmployee() {
        return employeeRepository.findAll();
    }

    public Mono<Employee> getEmployeeById(Long id) {
        return employeeRepository.findById(id);
    }

    public Mono<Void> deleteEmployeeById(Long id) {
        return employeeRepository.deleteById(id);
    }
}
```

4. In the following REST controller class, you can see that all the endpoints either return Flux or Mono reactive types:

```
import com.example.demo.model.Employee;
import com.example.demo.service.EmployeeService;
import org.springframework.http.ResponseEntity;
import org.springframework.web.bind.annotation.*;
import reactor.core.publisher.Flux;
import reactor.core.publisher.Mono;
```

```java
@RestController
@RequestMapping("/employee")
public class EmployeeController {

    private final EmployeeService employeeService;

    public EmployeeController(EmployeeService
      employeeService) {
        this.employeeService = employeeService;
    }

    @GetMapping
    public Flux<Employee> getAllEmployee() {
        return employeeService.getAllEmployee();
    }

    @PostMapping
    public Mono<Employee> createEmployee(@RequestBody
      Employee employee) {
        return
          employeeService.createEmployee(employee);
    }

    @GetMapping("/{id}")
    public Mono<ResponseEntity<Employee>>
      getEmployee(@PathVariable Long id) {
        Mono<Employee> employee =
          employeeService.getEmployeeById(id);
        return employee.map(e -> ResponseEntity.ok(e))
          .defaultIfEmpty(ResponseEntity.
            notFound().build());
    }

    @DeleteMapping("/{id}")
    public Mono<ResponseEntity<Void>>
      deleteUserById(@PathVariable Long id) {
```

```
            return employeeService.deleteEmployeeById(id)
                .map(r ResponseEntity.ok().
                  <Void>build())
                .defaultIfEmpty(ResponseEntity.notFound()
    .             build());
    }
}
```

The following is the Employee domain class:

```
public class Employee {
    @Id
    private Long id;
    private String firstName;
    private String lastName;
    private int age;
    private double salary;

    public Employee(String firstName, String lastName,
      int age, double salary) {
        this.firstName = firstName;
        this.lastName = lastName;
        this.age = age;
        this.salary = salary;
    }
    public Long getId() {
        return id;
    }

    public void setId(Long id) {
        this.id = id;
    }

    public String getFirstName() {
        return firstName;
    }

    public void setFirstName(String firstName) {
```

```java
        this.firstName = firstName;
    }

    public String getLastName() {
        return lastName;
    }

    public void setLastName(String lastName) {
        this.lastName = lastName;
    }

    public int getAge() {
        return age;
    }

    public void setAge(int age) {
        this.age = age;
    }

    public double getSalary() {
        return salary;
    }

    public void setSalary(double salary) {
        this.salary = salary;
    }
}
```

5. Let's run this application with the `mvn spring-boot:run` command. You will see the following logs once the application is up and running:

```
2021-07-13 20:40:12.979  INFO 47848 --- [          main]
com.example.demo.ReactiveApplication    : No active
profile set, falling back to default profiles: default
2021-07-13 20:40:14.268  INFO 47848 --- [          main]
.s.d.r.c.RepositoryConfigurationDelegate : Bootstrapping
Spring Data R2DBC repositories in DEFAULT mode.
2021-07-13 20:40:14.379  INFO 47848 --- [          main]
```

```
.s.d.r.c.RepositoryConfigurationDelegate : Finished
Spring Data repository scanning in 102 ms. Found 1 R2DBC
repository interfaces.

2021-07-13 20:40:17.627  INFO 47848 --- [          main]
o.s.b.web.embedded.netty.NettyWebServer  : Netty started
on port 8080

2021-07-13 20:40:17.652  INFO 47848 --- [          main]
com.example.demo.ReactiveApplication     : Started
ReactiveApplication in 5.889 seconds (JVM running for
7.979)

2021-07-13 20:40:17.921  INFO 47848 --- [          main]
com.example.demo.ReactiveApplication     : Saving
employee reactor.core.publisher.LambdaSubscriber@7dee835
```

The following is the output after accessing the /employee REST endpoint:

```
[{"id":1,"firstName":"Peter","lastName":"Parker","age":25,"salary":20000.0},
{"id":2,"firstName":"Tony","lastName":"Stark","age":30,"salary":40000.0},
{"id":3,"firstName":"Clark","lastName":"Kent","age":31,"salary":60000.0},
{"id":4,"firstName":"Bruce","lastName":"Wayne","age":33,"salary":100000.0}]
```

Figure 6.2 – REST endpoint response

In this section, we learned about the reactive programming model and created a Reactive Spring Boot CRUD application. In the next section, we will look at the different ways of containerizing your Java application with Skaffold.

Working with Skaffold container image builders

From *Chapter 3, Skaffold – Easy-Peasy Cloud-Native Kubernetes Application Development*, we know that Skaffold currently supports the following container image builders:

- Dockerfile
- Jib (Maven and Gradle)
- Bazel
- Cloud-native Buildpacks
- Custom scripts
- kaniko
- Google Cloud Build

In this section, we will cover these in detail by using them with the Spring Boot application we just built in the previous section. Let's talk about Dockerfile first.

Dockerfile

Docker is the gold standard for creating containers for many years. Even though there are many alternatives to Docker today, but it is still alive and kicking. Docker architecture depends on a daemon process that must be running to service all of your Docker commands. Then there is a Docker CLI that sent the commands to the Docker daemon for execution. The daemon process does what is required to push, pull, run container images, and so on. Docker expects a file called Dockerfile, handwritten by you, containing steps and instructions that it understands. This Dockerfile is then used to create a container image of your application with a command such as `docker build`. The advantage here is that this allows for a different customization level while making a container image of your application, as per your needs.

To build an image with Docker, we need to add a few instructions to our Dockerfile. Those instructions act as input, and then the Docker daemon process creates an image with those instructions. Let's look at an example to understand the working of a typical Dockerfile for a Java application.

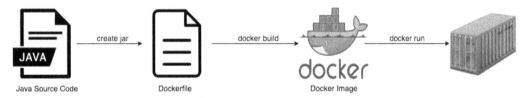

Figure 6.3 – Docker build flow

We will use the following Dockerfile to containerize our application:

```
FROM openjdk:16
COPY target/*.jar app.jar
ENTRYPOINT ["java","-jar","/app.jar"]
```

From the preceding code block, we can see the following:

- The `FROM` instruction indicates the base image for our application.

- The `COPY` instruction, as the name suggests, will copy the local .jar file that was built by Maven into our image.

- The `ENTRYPOINT` instruction acts as an executable for our container while it's starting.

In the `skaffold.yaml` file, we have added a new profile named `docker`. The following are the relevant parts of the `docker` profile:

```
profiles:
  - name: docker
    build:
      artifacts:
        - image: reactive-web-app
      local: {}
```

We can run the build with the `skaffold dev -profile=docker` command. The output should be similar to what we saw previously in *Figure 6.2*.

Jib

Jib (`https://github.com/GoogleContainerTools/jib`) stands for **Java Image Builder** and is purely written in Java. You already know that Jib allows Java developers to build containers using build tools such as Maven and Gradle. However, it has a CLI tool that can be used for non-Java applications such as Python on Node.js.

The significant advantage of using Jib is that you don't need to know anything about installing Docker or maintaining a Dockerfile. To containerize your Java application, you don't have to go through countless Docker tutorials. Jib is daemonless. Furthermore, as Java developers, we only care about the artifact (that is, jar file), and with Jib, we don't have to deal with any of the Docker commands. Using Jib, a Java developer can add the plugin to the build tool of their choice (Maven/Gradle), and with minimum configuration, you have your application containerized. Jib takes your application source code as input and outputs the container image of your application. Following is the build flow of your Java application with Jib:

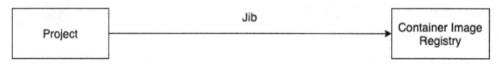

Figure 6.4 – Jib build flow

Let's try to build the application we created in the previous section with Jib:

1. First, we will create the skaffold.yaml file using the Skaffold init command, as follows:

```yaml
apiVersion: skaffold/v2beta20
kind: Config
metadata:
  name: reactive-web-app
build:
  artifacts:
  - image: reactive-web-app
    jib:
      fromImage: adoptopenjdk:16-jre
      project: com.example:reactive-web-app
      args:
        - -DskipTests
deploy:
  kubectl:
    manifests:
    - k8s/manifest.yaml
```

> **Tip**
>
> Jib cleverly split your application image layers into the following for faster rebuilds:
>
> -Classes
>
> -Resources
>
> -Project Dependencies
>
> -Snapshot and all other dependencies
>
> The goal is to separate files that often change as compared to the ones which are rarely changed. The immediate benefit is that you don't have to rebuild the entire application because Jib only rebuilds the layer containing the changed files and reuse cached layers for files that didn't change.
>
> With Jib, you might see the following warning in logs if you don't specify the image digest:
>
> [WARNING] Base image `'adoptopenjdk/openjdk16'` does not use a specific image `digest` - `build` may not be reproducible.
>
> You can overcome this by using the proper image digest. For example, in `maven-jib-plugin`, you can make the following changes, while in the `skaffold.yaml` file, you can specify the image digest:

```
<plugin>

    <groupId>com.google.cloud.tools</groupId>

    <artifactId>jib-maven-plugin</artifactId>

    <version>3.1.1</version>

    <configuration>

        <from>

            <image>adoptopenjdk/openjdk16@
                sha256:b40f81a9f7e7e4533ed0c
                6ac794ded9f653807f757e2b8b4e1
                fe729b6065f7f5</image>

        </from>

        <to>

            <image>docker.io/hiashish/image</image>

        </to>

    </configuration>

</plugin>
```

Following is the Kubernetes service manifest:

```
apiVersion: v1
kind: Service
metadata:
  name: reactive-web-app
spec:
  ports:
    - port: 8080
      protocol: TCP
      targetPort: 8080
  type: Loadbalancer
  selector:
    app: reactive-web-app
```

Following is the Kubernetes deployment manifest:

```
apiVersion: apps/v1
kind: Deployment
metadata:
  name: reactive-web-app
spec:
  selector:
    matchLabels:
      app: reactive-web-app
  template:
    metadata:
      labels:
        app: reactive-web-app
    spec:
      containers:
        - name: reactive-web-app
          image: reactive-web-app
```

2. Now, we must run the `skaffold dev` command. The following is the output:

```
skaffold dev
Listing files to watch...
 - reactive-web-app
Generating tags...
 - reactive-web-app -> reactive-web-app:fcda757-dirty
Checking cache...
 - reactive-web-app: Found Locally
Starting test...
Tags used in deployment:
 - reactive-web-app -> reactive-web-app:3ad471bdebe8e0606
040300c9b7f1af4bf6d0a9d014d7cb62d7ac7b884dcf008
Starting deploy...
 - service/reactive-web-app created
 - deployment.apps/reactive-web-app created
Waiting for deployments to stabilize...
 - deployment/reactive-web-app is ready.
Deployments stabilized in 3.34 seconds
Press Ctrl+C to exit
Watching for changes...
```

With minikube, we can use the `minikube service reactive-web-app` command to open the exposed service. We will use the URL mentioned in the following screenshot to access our application:

```
|-----------|-------------------|---------------|------------------------------|
| NAMESPACE |       NAME        |  TARGET PORT  |            URL               |
|-----------|-------------------|---------------|------------------------------|
| default   | reactive-web-app  |         8080  | http://192.168.49.2:31485    |
|-----------|-------------------|---------------|------------------------------|
   Starting tunnel for service reactive-web-app.
|-----------|-------------------|---------------|------------------------------|
| NAMESPACE |       NAME        |  TARGET PORT  |            URL               |
|-----------|-------------------|---------------|------------------------------|
| default   | reactive-web-app  |               | http://127.0.0.1:55174       |
|-----------|-------------------|---------------|------------------------------|
   Opening service default/reactive-web-app in default browser...
 ! Because you are using a Docker driver on darwin, the terminal needs to be open to run it.
```

Figure 6.5 – Exposed service URL

After accessing the `http://127.0.0.1:55174/employee` URL, we should get an output similar to *Figure 6.2*.

Bazel

Bazel is an open source, multilanguage, fast, and scalable build tool similar to Maven and Gradle. Skaffold provides support for Bazel and it can load images to the local Docker daemon. Bazel requires two files: `WORKSPACE` and `BUILD`.

The `WORKSPACE` file is typically available at the root directory of your project. This file indicates the Bazel workspace. It looks for build inputs and stores the build output in the directory where the `WORKSPACE` file was created.

The `BUILD` file instructs Bazel on what to build and how to build different parts of your project. The following is an example of a `BUILD` file for a Java application. In this example, we are instructing Bazel to use the `java_binary` rule to create a `.jar` file for our application:

```
java_binary(
name = "ReactiveWebApp",
srcs = glob(["src/main/java/com/example/*.java"]),)
```

To build your project, you can run commands such as `build //: ReactiveWebApp`. The following is the `skaffold.yaml` file, which contains a `bazel` profile:

```
profiles:
  - name: bazel
    build:
      artifacts:
        - image: reactive-web-app
          bazel:
            target: //:reactive-web-app.tar
```

Next we have Buildpacks.

Buildpacks

Heroku first created Buildpacks (`https://buildpacks.io/`) in 2011. It is now part of the CNCF foundation. Just like Jib, Buildpacks can also work without the need for a Dockerfile. However, you will need a Docker daemon process up and running for it to work. With Buildpacks, the input is your application source code, and the output is the container image. It's pretty similar to Jib in this respect, though Jib can work without a Docker daemon.

In the background, Buildpacks does a lot of work, including retrieving dependencies, processing assets, handling caching, and compiling code for whatever language your app has been built in:

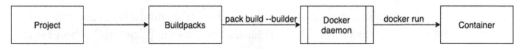

Figure 6.6 – Buildpacks build flow

As explained earlier, Skaffold requires a local Docker daemon to build an image with Buildpacks. Skaffold will execute the build inside a container using the builder specified in the Buildpacks configuration of your `skaffold.yaml` file. Also, you don't have to install the pack CLI as the Google Cloud Buildpacks project (`https://github.com/GoogleCloudPlatform/buildpacks`) provides builder images for tools such as Skaffold. You can choose to skip this, but upon successfully building it, Skaffold will push the image to the remote registry.

> **Tip**
>
> Starting with Spring Boot version 2.3, Spring Boot includes direct Buildpacks support for both Maven and Gradle projects. With the `mvn spring-boot:build-image` command, you can create an image of your application that's loaded to your locally running Docker daemon. Although you are not required to maintain a Dockerfile, Buildpacks depends on the Docker daemon process. If you don't have the Docker daemon running locally, you will get the following error while executing the Maven command:
>
> ```
> Failed to execute goal org.springframework.
> boot:spring-boot-maven-plugin:2.4.2:build-
> image (default-cli) on project imagebuilder:
> Execution default-cli of goal org.springframework.
> boot:spring-boot-maven-plugin:2.4.2:build-
> image failed: Connection to the Docker daemon at
> 'localhost' failed with error "[61] Connection
> refused"; ensure the Docker daemon is running and
> accessible
> ```

To build our application with Buildpacks, we have added a new profile `name` of `pack` and used that to add a new `build` section to the `skaffold.yaml` configuration file. In the `builder` field, we are instructing Skaffold to use the `gcr.io/buildpacks/builder:v1` builder image. The following are the relevant parts of the configuration file:

```
profiles:
  - name: pack
    build:
      artifacts:
        - image: reactive-web-app
          buildpacks:
            builder: gcr.io/buildpacks/builder:v1
            env:
              - GOOGLE_RUNTIME_VERSION=16
```

We can run the build with the `skaffold dev -profile=pack` command. The output should be similar to what we saw in *Figure 6.2*.

> **Tip**
> Spring Boot Buildpacks integration can be used to push an image to a remote container registry. We would need to make the following changes in `pom.xml` to do this:

```
<plugin>
    <groupId>org.springframework.boot</groupId>
      <artifactId>spring-boot-maven-plugin</artifactId>
    <configuration>
        <image>
            <name>docker.example.com/library/$
            {project.artifactId}</name>
            <publish>true</publish>
        </image>
        <docker>
          <publishRegistry>
            <username>user</username>
            <password>secret</password>
            <url>https://docker.example.com/v1/</url>
            <email>user@example.com</email>
```

```
        </publishRegistry>
      </docker>
    </configuration>
  </plugin>
```

Custom script

You can use the custom script option if none of the supported container image builders work for your use case. With this option, you can write custom scripts or choose a build tool of your liking. You can configure a custom script by adding a custom field to each corresponding artifact in the build section of the skaffold.yaml file.

In the following example skaffold.yaml file, we have created a new profile named custom. In the buildCommand field, we have used the build.sh script to containerize our Spring Boot application:

```
profiles:
  - name: custom
    build:
      artifacts:
        - image: reactive-web-app
          custom:
            buildCommand: sh build.sh
```

The build.sh script file contains the following content. It uses the docker build command to create an image of our application. Skaffold will supply $IMAGE (that is, the fully qualified image name environment variable) to the custom build script:

```
#!/bin/sh
set -e
docker build -t "$IMAGE" .
```

Next we move to kaniko.

kaniko

kaniko is an open source tool that's used to build container images from a Dockerfile inside a container or Kubernetes cluster. kaniko doesn't require privileged root access to build container images.

kaniko has no dependency on a Docker daemon and executes each command within a Dockerfile entirely in the user space. With kaniko, you can start building container images in environments that can't securely run a Docker daemon, such as a standard Kubernetes cluster. So, how does kaniko work? Well, kaniko uses an executor image called `gcr. io/kaniko-project/executor`, and this image runs inside a container. It is not recommended to run the kaniko executor binary in another image as it might not work.

Let's see how this is done:

1. We will use the following Dockerfile with kaniko to build the container image of our application:

    ```
    FROM maven:3-adoptopenjdk-16 as build
    RUN mkdir /app
    COPY . /app
    WORKDIR /app
    RUN mvn clean verify -DskipTests

    FROM adoptopenjdk:16-jre
    RUN mkdir /project
    COPY --from=build /app/target/*.jar /project/app.jar
    WORKDIR /project
    ENTRYPOINT ["java","-jar","app.jar"]
    ```

2. The following is the relevant part of `skaffold.yaml`:

    ```
    profiles:
      - name: kaniko
        build:
          cluster:
            pullSecretPath: /Users/ashish/Downloads/kaniko-
    secret.json
            artifacts:
              - image: reactive-web-app
                kaniko: {}
    ```

Here, we have added a new profile called `kaniko` to build our container images inside the Google Kubernetes cluster. You will learn more about GKE in *Chapter 8, Deploying a Spring Boot Application to Google Kubernetes Engine Using Skaffold*.

An important point to highlight in this `skaffold.yaml` file is that we would need a credential from the active Kubernetes cluster to build our image inside the cluster. For that, a GCP service account is required. This account has a storage admin role so that images can be pulled and pushed. We can use the following command to build and deploy our application to GKE:

```
skaffold run --profile=kaniko --default-repo=gcr.io/basic-
curve-316617
```

We will be using a remote Kubernetes cluster hosted on GCP for this demo.
Let's get started:

1. First, we need to create a service account for kaniko with permissions to pull and push images from/to `gcr.io`. Then, we need to download the JSON service account file and rename the file `kaniko-secret`. Also, make sure that you do not append `.json` to the filename; create a Kubernetes secret using the following command. You need to make sure that the Kubernetes context is set to a remote Kubernetes cluster:

    ```
    kubectl create secret generic kaniko-secret --from-
    file=kaniko-secret
    ```

2. Since we are going to push the image to **Google Container Registry (GCR)**, we have mentioned the `-default-repo` flag so that it always points to GCR. The following are the logs:

    ```
    Generating tags...
     - reactive-web-app -> gcr.io/basic-curve-316617/
    reactive-web-app:fcda757-dirty
    Checking cache...
     - reactive-web-app: Not found. Building
    Starting build...
    Checking for kaniko secret [default/kaniko-secret]...
    Creating kaniko secret [default/kaniko-secret]...
    Building [reactive-web-app]...
    INFO[0000] GET KEYCHAIN
    INFO[0000] running on kubernetes ....
    ```

In the following logs, you can see that kaniko started building images inside the container by downloading base images for different stages of the build. kaniko started packaging and downloading the dependencies for our Spring Boot application:

```
INFO[0001] Retrieving image manifest adoptopenjdk:16-jre
INFO[0001] Retrieving image adoptopenjdk:16-jre from
registry index.docker.io
INFO[0001] GET KEYCHAIN
INFO[0001] Built cross stage deps: map[0:[/app/target/*.
jar]]
INFO[0001] Retrieving image manifest maven:3-
adoptopenjdk-16
. . . . . . . . . . . . . . .
INFO[0035] RUN mvn clean verify -DskipTests
INFO[0035] cmd: /bin/sh
INFO[0035] args: [-c mvn clean verify -DskipTests]
INFO[0035] Running: [/bin/sh -c mvn clean verify
-DskipTests]
[INFO] Scanning for projects...
Downloading from central: https://repo.maven.apache.org/
maven2/org/springframework/boot/spring-boot-starter-
parent/2.5.2/spring-boot-starter-parent-2.5.2.pom
```

3. In the following logs, you can see that the build was successful and that kaniko was able to push the image to GCR. Then, we deployed the image to the Google Kubernetes cluster using `kubectl`:

```
[INFO] BUILD SUCCESS
INFO[0109] Taking snapshot of full filesystem...
INFO[0114] Saving file app/target/reactive-web-app-0.0.1-
SNAPSHOT.jar for later use
. . . .
INFO[0130] COPY --from=build /app/target/*.jar /project/
app.jar
. . . .
INFO[0131] ENTRYPOINT ["java","-jar","app.jar"]
INFO[0131] GET KEYCHAIN
INFO[0131] Pushing image to gcr.io/basic-curve-316617/
reactive-web-app:fcda757-dirty
INFO[0133] Pushed image to 1 destinations
```

```
Starting test...
Tags used in deployment:
  - reactive-web-app -> gcr.io/basic-curve-316617/
reactive-web-app:fcda757-dirty@9797e8467bd25fa4a237
e21656cd574c0c46501e5b3233a1f27639cb5b66132e
Starting deploy...
  - service/reactive-web-app created
  - deployment.apps/reactive-web-app created
Waiting for deployments to stabilize...
  - deployment/reactive-web-app: creating container
reactive-web-app
     - pod/reactive-web-app-6b885dcf95-q8dr5: creating
container reactive-web-app
  - deployment/reactive-web-app is ready.
Deployments stabilized in 12.854 seconds
```

In the following screenshot, we can see that after the deployment, a pod is running and that the service that's been exposed is of the **Load balancer** type:

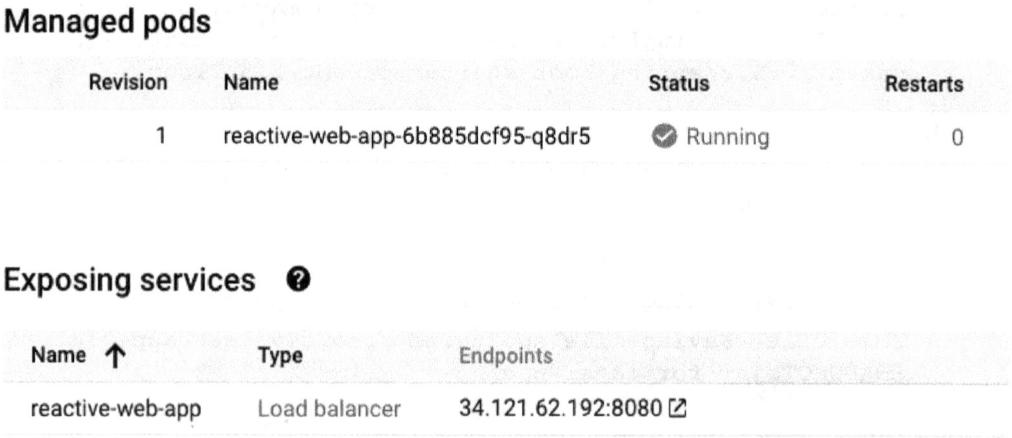

Managed pods

Revision	Name	Status	Restarts
1	reactive-web-app-6b885dcf95-q8dr5	✔ Running	0

Exposing services ❷

Name ↑	Type	Endpoints
reactive-web-app	Load balancer	34.121.62.192:8080 🗗

Figure 6.7 – Pod running and the service exposed for external access

The following is the output after accessing the /employee REST endpoint of our Spring Boot application using the endpoints exposed by GKE:

```
[{"id":1,"firstName":"Peter","lastName":"Parker","age":25,"salary":20000.0},
{"id":2,"firstName":"Tony","lastName":"Stark","age":30,"salary":40000.0},
{"id":3,"firstName":"Clark","lastName":"Kent","age":31,"salary":60000.0},
{"id":4,"firstName":"Bruce","lastName":"Wayne","age":33,"salary":100000.0}]
```

Figure 6.8 – REST application response

Google Cloud Build

Cloud Build is a service that runs your build using GCP infrastructure. Cloud Build works by importing source code from various repositories or Google Cloud Storage spaces, executing a build, and producing artifacts such as container images.

We created a new profile named gcb in skaffold.yaml to trigger the remote build of our application using Google Cloud Build. The following is the relevant part of the skaffold.yaml profile section:

```
profiles:
  - name: gcb
    build:
      artifacts:
        - image: reactive-web-app
          docker:
            cacheFrom:
              - reactive-web-app
      googleCloudBuild: {}
```

We can run the following command to start the remote build of our application with Google Cloud Build:

```
skaffold run --profile=gcb --default-repo=gcr.io/basic-
curve-316617
```

If this is your first time doing this, make sure that you have enabled the Cloud Build API, either from the **Cloud Console** dashboard or through the gcloud CLI. Otherwise, you may get the following error:

```
Generating tags...
  - reactive-web-app -> gcr.io/basic-curve-316617/reactive-web-
app:fcda757-dirty
```

```
Checking cache...
 - reactive-web-app: Not found. Building
Starting build...
Building [reactive-web-app]...
Pushing code to gs://basic-curve-316617_cloudbuild/source/
basic-curve-316617-046b951c-5062-4824-963b-a204302a77e1.tar.gz
could not create build: googleapi: Error 403: Cloud Build
API has not been used in project 205787228205 before or
it is disabled. Enable it by visiting https://console.
developers.google.com/apis/api/cloudbuild.googleapis.com/
overview?project=205787228205 then retry. If you enabled this
API recently, wait a few minutes for the action to propagate to
our systems and retry.
.....
```

You can enable the Cloud Build API via the **Cloud Console** dashboard by visiting the URL mentioned in the error logs and clicking on the **ENABLE** button, as shown in the following screenshot:

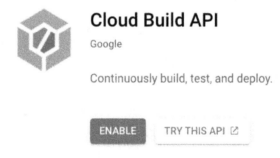

Figure 6.9 – Enabling the Cloud Build API

You need to make sure that, in your kubeconfig file, the GKE remote cluster is the active cluster for this deployment before running the actual command to start the build and deployment process. The following is the output of the skaffold run command. In the following logs, you can see that our entire source code is packaged as a tar.gz file and sent to the Google Cloud Storage location. From there, Cloud Build picks it and starts building our image:

```
skaffold run --profile=gcb --default-repo=gcr.io/basic-
```

```
curve-316617

Generating tags...

  - reactive-web-app -> gcr.io/basic-curve-316617/reactive-web-
app:fcda757-dirty

Checking cache...

  - reactive-web-app: Not found. Building

Starting build...

Building [reactive-web-app]...

Pushing code to gs://basic-curve-316617_cloudbuild/source/
basic-curve-316617-aac889cf-d854-4e7f-a3bc-b26ea06bf854.tar.gz

Logs are available at

https://console.cloud.google.com/m/cloudstorage/b/basic-
curve-316617_cloudbuild/o/log-43705458-0f75-4cfd-8532-
7f7db103818e.txt

starting build "43705458-0f75-4cfd-8532-7f7db103818e"

FETCHSOURCE

Fetching storage object: gs://basic-curve-316617_cloudbuild/
source/basic-curve-316617-aac889cf-d854-4e7f-a3bc-b26ea06bf854.
tar.gz#1626576177672677

Copying gs://basic-curve-316617_cloudbuild/source/basic-
curve-316617-aac889cf-d854-4e7f-a3bc-b26ea06bf854.tar.
gz#1626576177672677...

- [1 files][ 42.2 MiB/ 42.2 MiB]

Operation completed over 1 objects/42.2 MiB.

BUILD

Starting Step #0

Step #0: Already have image (with digest): gcr.io/cloud-
builders/docker

...
```

In the following logs, you can see that the image has been built, tagged, and pushed to GCR. Then, using kubectl, the application is deployed to GKE, as follows:

```
Step #1: Successfully built 1a2c04528dad

Step #1: Successfully tagged gcr.io/basic-curve-316617/
reactive-web-app:fcda757-dirty

Finished Step #1

PUSH

Pushing gcr.io/basic-curve-316617/reactive-web-app:fcda757-
```

```
dirty
The push refers to repository [gcr.io/basic-curve-316617/
reactive-web-app]
7a831de44071: Preparing
574a11c0c1c8: Preparing
783bfc5acd81: Preparing
2da4fab53cd6: Preparing
a70daca533d0: Preparing
783bfc5acd81: Layer already exists
2da4fab53cd6: Layer already exists
a70daca533d0: Layer already exists
574a11c0c1c8: Pushed
7a831de44071: Pushed
fcda757-dirty: digest: sha256:22b2de72d3e9551f2531f2b9dcdf5e-
4b2eabaabc9d1c7a5930bcf226e6b9c04b size: 1372
DONE
Starting test...
Tags used in deployment:
 - reactive-web-app -> gcr.io/basic-curve-316617/reactive-web-
app:fcda757-dirty@sha256:22b2de72d3e9551f2531f2b9dcdf5e4b2
eabaabc9d1c7a5930bcf226e6b9c04b
Starting deploy...
 - service/reactive-web-app configured
 - deployment.apps/reactive-web-app created
Waiting for deployments to stabilize...
 - deployment/reactive-web-app: creating container reactive-
web-app
    - pod/reactive-web-app-789f775d4-z998t: creating container
reactive-web-app
 - deployment/reactive-web-app is ready.
Deployments stabilized in 1 minute 51.872 seconds
```

In the **Workload** section of GKE, you can see that **reactive-web-app** has been deployed and that its status is OK, as follows:

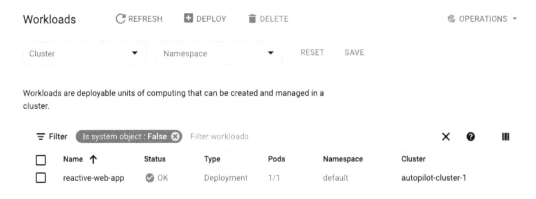

Figure 6.10 – Application deployed successfully on GKE

In this section, we learned about the different ways we can containerize our Reactive Spring Boot CRUD application.

In the next section, we will explore different ways to deploy an application to Kubernetes with Skaffold.

Exploring Skaffold container image deployers

In this section, we will look at the container image deployment methods supported by Skaffold. With Skaffold, you can deploy your application to Kubernetes using the following three tools:

- Helm
- kubectl
- Kustomize

Let's discuss them in detail.

Helm

Helm is the package manager, and **charts** are the packages for your Kubernetes applications. It allows you to define, install, and update your Kubernetes application easily. You can write charts for your applications or use production-ready, pre-packaged charts for popular software such as MySQL and MongoDB from a stable chart repository.

Until Helm 2, Helm followed a client-server architecture. However, due to significant changes being made to the architecture with Helm 3, it is a client-only architecture. Therefore, there is no need to have a server-side component such as **Tiller** installed on your Kubernetes cluster.

Now, let's learn more about Helm:

1. Skaffold will not install Helm for us, so we must install it using the Homebrew package manager for macOS:

    ```
    $ brew install helm
    ```
    ```
    $ helm version
    ```
    ```
    version.BuildInfo{Version:"v3.6.3",
    GitCommit:"d506314abfb5d21419df8c7e7e68012379db2354",
    GitTreeState:"dirty", GoVersion:"go1.16.5"}
    ```

    For Windows, you can download it using chocolatey:

    ```
    choco install kubernetes-helm
    ```

    You can also download Helm using an installer script, which will download the latest version:

    ```
    $ curl -fsSL -o get_helm.sh https://raw.
    githubusercontent.com/helm/helm/master/scripts/get-helm-3
    ```
    ```
    $ chmod 700 get_helm.sh
    ```
    ```
    $ ./get_helm.sh
    ```

2. Next, we will create a Helm chart skeleton using the following command:

    ```
    $ helm create reactive-web-app-helm
    Creating charts
    ```

3. We will create a new Skaffold profile called `jibWithHelm` to build an image with Jib and then deploy it using Helm:

```
profiles:
    - name: jibWithHelm
    build:
      artifacts:
        - image: gcr.io/basic-curve-316617/reactive-
          web-app-helm
          jib:
            args:
              - -DskipTests
    deploy:
      helm:
        releases:
          - name: reactive-web-app-helm
            chartPath: reactive-web-app-helm
            artifactOverrides:
              imageKey: gcr.io/basic-curve-
                316617/reactive-web-app-helm
            valuesFiles:
              - reactive-web-app-helm/values.yaml
            imageStrategy:
              helm: { }
```

Here, the image name under the `build` section should match the image name given under the `artifactOverrides` section of the `skaffold.yaml` file. Otherwise, you will get an error.

We have also provided the path to the `values.yaml` file under the `valuesFiles` section of the `skaffold.yaml` file.

The typical convention for defining image references with Helm is through the `values.yaml` file. The following is the content of the `values.yaml` file that will be referenced by Helm:

```
replicaCount: 1
imageKey:
  repository: gcr.io/basic-curve-316617
  pullPolicy: IfNotPresent
```

```
    tag: latest
service:
    type: LoadBalancer
    port: 8080
    targetPort: 8080
```

The values in the `values.yaml` file will be referenced inside a templated resource file, as shown in the following code snippet. This templated file is located inside `reactive-web-app-helm/templates/**.yaml`:

```
    spec:
        containers:
            - name: {{ .Chart.Name }}
                image: {{ .Values.imageKey.repository }}:{{
                    .Values.imageKey.tag }}
                imagePullPolicy: {{ .Values.imageKey.pullPolicy }}
```

After running `skaffold run --profile=jibWithHelm`, Skaffold will build the image using Jib and deploy it to GKE using Helm charts. This will result in the following output:

```
skaffold run --profile=jibWithHelm
Generating tags...
 - gcr.io/basic-curve-316617/reactive-web-app-helm -> gcr.io/
basic-curve-316617/reactive-web-app-helm:3ab62c6-dirty
Checking cache...
 - gcr.io/basic-curve-316617/reactive-web-app-helm: Found
Remotely
Starting test...
Tags used in deployment:
 - gcr.io/basic-curve-316617/reactive-web-app-
helm -> gcr.io/basic-curve-316617/reactive-web-app-
helm:3ab62c6-dirty@sha256:2d9539eb23bd9db578feae7e4956c-
30d9320786217a7307e0366d9cc5ce359bc
Starting deploy...
Helm release reactive-web-app-helm not installed. Installing...
NAME: reactive-web-app-helm
LAST DEPLOYED: Thu Aug 26 11:34:39 2021
NAMESPACE: default
```

```
STATUS: deployed
REVISION: 1
Waiting for deployments to stabilize...
 - deployment/reactive-web-app-helm is ready.
Deployments stabilized in 3.535 seconds
```

We can verify whether the pods are running by going to the **Workloads** section of GKE. In the following screenshot, we can see that we have a pod running:

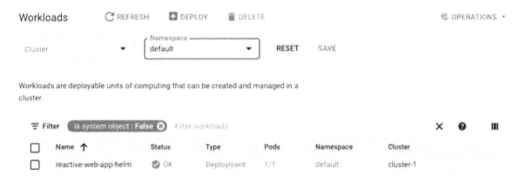

Figure 6.11 – Helm Charts deployed successfully on GKE

Similarly, under the **Services & Ingress** section, we can see that an **External load balancer** type of service has been exposed for external access:

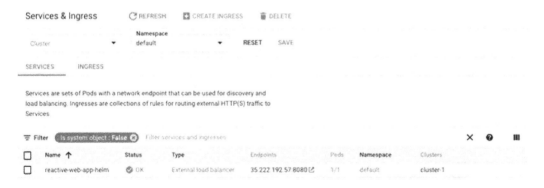

Figure 6.12 – LoadBalancer service type exposed on GKE

After accessing the application using the URL mentioned in the **Endpoints** column, the output should be similar to what we saw in *Figure 6.2*.

kubectl

kubectl is a command-line tool that's used to run commands on your Kubernetes cluster. It interacts with the Kubernetes API Server to run those commands. You can use it to accomplish various tasks, such as viewing logs of a pod, create Kubernetes

resources such as deployments, find out about the state of your cluster, and pods among others. In the following code snippet, you can see that we are using kubectl for deployment purposes. The Kubernetes manifests are under the `k8s` directory:

```
deploy:
  kubectl:
    manifests:
    - k8s/manifest.yaml
```

Kustomize

Kustomize, as its name suggests, is a template-free declarative approach to Kubernetes configuration, management, and customization options. With Kustomize, we provide a base skeleton and patches. In this approach, compared to Helm, we provide a base deployment and then describe the differences for different environments. For example, we can have different numbers of replicas and health checks for production as compared to staging. Kustomize can installed separately and since version 1.14 of kubectl, we can use the `-k` command. Follow the instructions mentioned at `https://kubectl.docs.kubernetes.io/installation/kustomize/` to install it on your supported OS.

In the following example, we have a profile called `kustomizeProd` and are using Kustomize as the deployment strategy for our application:

```
profiles:
- name: kustomizeProd
  build:
    artifacts:
      - image: reactive-web-app
        jib:
          args:
            - -DskipTests
  deploy:
    kustomize:
      paths:
        - kustomization/overlays/prod
```

We must have the following directory structure for Kustomize to work properly. In the following directory, you can see that under the `kustomization/base` directory, we have our original YAML files describing the resources we want to deploy in the GKE cluster. We will never touch these files; instead, we will just apply customization above them to create new resources definitions:

```
├── kustomization
│   ├── base
│   │   ├── deployment.yaml
│   │   ├── kustomization.yaml
│   │   └── service.yaml
│   └── overlays
│       ├── dev
│       │   ├── environment.yaml
│       │   └── kustomization.yaml
│       └── prod
│           ├── increase_replica.yaml
│           ├── kustomization.yaml
│           └── resources_constraint.yaml
```

We have a file named `kustomization.yaml` inside this `base` folder. It describes the resources you use. The resources are the paths to the Kubernetes manifests files relative to the current file:

```
apiVersion: kustomize.config.k8s.io/v1beta1
kind: Kustomization
resources:
  - deployment.yaml
  - service.yaml
```

Next, we have the `kustomization/overlays/prod` folder, which has a `kustomization.yaml` file inside it. It contains the following content:

```
apiVersion: kustomize.config.k8s.io/v1beta1
kind: Kustomization
resources:
  - ../../base
patchesStrategicMerge:
  - increase_replica.yaml
  - resources_constraint.yaml
```

If you can see, in `base`, we didn't define any environment variables, a replicas count, or resource constraints. But for production scenarios, we must add those things above our base. To do so, we just have to create the chunk of YAML we would like to apply above our base and reference it inside the `kustomization.yaml` file. We have already added this YAML to the list of `patchesStrategicMerge` in the `kustomization.yaml` file.

The `increase_replica.yaml` file contains two replicas and looks like this:

```
apiVersion: apps/v1
kind: Deployment
metadata:
  name: reactive-web-app
spec:
  replicas: 2
```

The `resources_constraint.yaml` file contains the resource request and limit and looks like this:

```
apiVersion: apps/v1
kind: Deployment
metadata:
  name: reactive-web-app
spec:
  template:
    spec:
      containers:
        - name: reactive-web-app
          resources:
            requests:
              memory: 512Mi
              cpu: 256m
            limits:
              memory: 1Gi
              cpu: 512m
```

Now, we can run the `skaffold run --profile=kustomizeProd --default-repo=gcr.io/basic-curve-316617` command. This will deploy the application to GKE using Kustomize. The output we get should be similar to what we saw previously in *Figure 6.2*.

In this section, we looked at the tools we can use with Skaffold to deploy applications to a Kubernetes cluster.

Summary

In this chapter, we started by introducing ourselves to reactive programming and built a Spring Boot CRUD application. We were also introduced to Skaffold's supported container image builders, including Docker, kaniko, Jib, and Buildpacks. We covered them by looking at their practical implementations. We also discussed the different ways we can deploy images to a Kubernetes cluster using tools such as kubectl, Helm, and Kustomize.

In this chapter, we gained a solid understanding of tools such as Jib, kaniko, Helm, and Kustomize, to name a few. You can apply your knowledge of these tools to build and deploy your containers.

In the next chapter, we will build and deploy a Spring Boot application to Kubernetes using Google's Cloud Code extension.

Further reading

To learn more about Skaffold, take a look at the Skaffold documentation: `https://skaffold.dev/docs/`.

Section 3: Building and Deploying Cloud-Native Spring Boot Applications with Skaffold

This section will primarily focus on building and deploying Spring Boot applications using Skaffold to local (minikube and so on) and remote clusters (**GKE**). We will explore how you can build and deploy cloud-native applications from the comfort of your IDE using Cloud Code developed by Google. Then we will build and deploy a Spring Boot application to a managed Kubernetes offering such as **GKE** using Skaffold. We will further learn about creating a production-ready CI/CD pipeline using Skaffold and GitHub Actions. We will do some experiments by combining Skaffold and Argo CD to implement a GitOps-style CD workflow. Finally, we will look at some alternatives to Skaffold and learn about Skaffold best practices that we should utilize in our workflow. Furthermore, we will explore the most common pitfalls/limitations of developing an application with Skaffold. Finally, we will summarize what we have learned throughout this book.

In this section, we have the following chapters:

- *Chapter 7, Building and Deploying a Spring Boot Application with the Cloud Code Plugin*
- *Chapter 8, Deploying a Spring Boot Application to Google Kubernetes Engine Using Skaffold*
- *Chapter 9, Creating a Production-Ready CI/CD Pipeline with Skaffold*
- *Chapter 10, Exploring Skaffold Alternatives, Best Practices, and Pitfalls*

7
Building and Deploying a Spring Boot Application with the Cloud Code Plugin

In the previous chapter, we learned about Skaffold-supported container image builders and deployers. In this chapter, we will introduce you to Google's Cloud Code plugin, which is available with IDEs such as IntelliJ. We will create a Spring Boot application and use the Cloud Code plugin to deploy it to a local Kubernetes cluster.

In this chapter, we're going to cover the following main topics:

- Introducing Google's Cloud Code plugin
- Installing and working with the IntelliJ Cloud Code plugin
- Creating a Spring Boot application
- Containerizing and deploying a Spring Boot application using Cloud Code

By the end of this chapter, you will have a solid understanding of the Cloud Code plugin and how you can use it to accelerate the development life cycle of a Kubernetes application using an IDE.

Technical requirements

For this chapter, you will need the following:

- Visual Studio Code (`https://code.visualstudio.com/download`) or the IntelliJ IDE (`https://www.jetbrains.com/idea/download/`)
- Git
- Spring Boot 2.5
- OpenJDK 16

The code from this book's GitHub repository can be found at `https://github.com/PacktPublishing/Effortless-Cloud-Native-App-Development-Using-Skaffold/tree/main/Chapter07`.

Introducing Google's Cloud Code plugin

If you are working to develop or maintain cloud-native applications in today's age, then it is a sort of unspoken truth that you need a set of tools or a tool to ease your development process. As developers, we typically do the following tasks in the inner development loop:

- Download specific dependencies such as Skaffold, `minikube`, and `kubectl` to set up the local development environment.

- Do a lot of context switching to view logs, documentation, and browse the cloud vendor-provided console.

While Skaffold is an excellent solution to this problem, would it not be nice to have everything clubbed into your IDE? For example, we can add a plugin that can do all these tasks and focus on the coding part. For this, we can use the **Google Cloud Code** extension, since it simplifies the development of cloud-based applications with your favorite IDE, such as IntelliJ, Visual Studio Code, and so on.

Let's understand some of the features that Cloud Code offers:

- Write, debug, and deploy Kubernetes application faster.
- Support for multiple IDEs, including JetBrains IntelliJ, Visual Studio Code, and Cloud Shell Editor.
- Multiple startup templates for different languages with best practices to start your development in no time.
- You can deploy your application with a single click on Google Kubernetes Engine or Cloud Run.
- Efficiently works with other Google Cloud Platform services, including Google Kubernetes Engine, Google Container Registry, and Cloud Storage.
- Improves the YAML file editing process with features such as code snippets and in-line documentation.
- Built-in support for Skaffold to fasten your inner development loop.
- Easy remote and local debugging of your applications running on Kubernetes.
- Built-in log viewer to view application logs in real time for your Kubernetes applications.

Now that we have understood what Cloud Code is and its features, let's try to install and use its startup templates to quickly deploy a Java application to the local Kubernetes cluster.

Installing and working with the IntelliJ Cloud Code plugin

To get started with the Cloud Code plugin, first, we need to download it. You can access the IntelliJ plugin marketplace to download it. Let's learn how to do this:

1. For Windows or Linux, navigate to **File | Settings | Plugins**, enter **Cloud Code** in the search area, and click on **Install**.

2. For macOS, navigate to **IntelliJ IDEA | Preferences | Plugins**, enter **Cloud Code** in the search area, and click on **Install**, as shown in the following screenshot:

Figure 7.1 – Installing Cloud Code from the IntelliJ marketplace

3. Once the download is completed, a welcome screen will pop up. Here, click on **Create a Kubernetes Sample App**, as shown in the following screenshot:

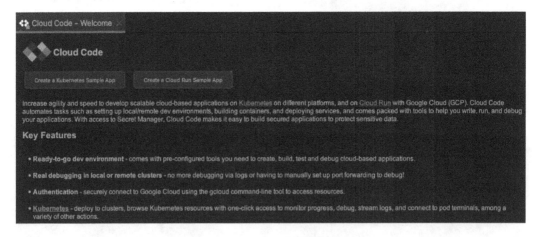

Figure 7.2 – Cloud Code welcome page

4.   On the next screen, a **New Project** window will open. We need to select the **Java: Guestbook** project, as shown in the following screenshot, and click **Next**:

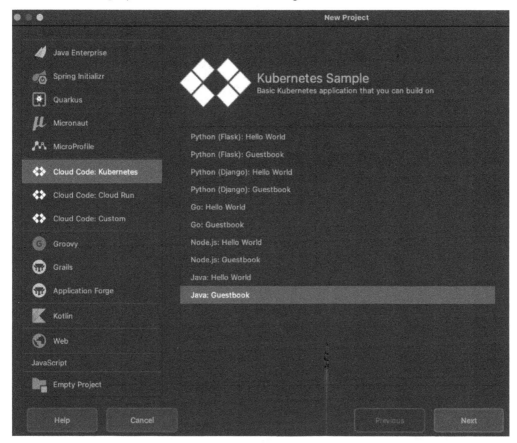

Figure 7.3 – Selecting a pre-built Java Guestbook application

5.   On the next screen, you will be asked to specify your container image repository. If you are using DockerHub, GCR, or any other image registry, then add those details and click on **Next**. For example, if you are using GCR, then enter something like `gcr.io/gcp-project-id`. Since we are using startup templates and the image name is already defined in Kubernetes manifests, we can leave that part.

6.   On the next screen, enter the project name and click on **Finish**. The sample Java project will be downloaded to your default project location.

7.   Now that we have a working project, click the **Run/Debug Configurations** dropdown and select **Edit Configurations**.

8. In the **Run/Debug Configurations** dialog box, select the **Develop on Kubernetes** configuration. Then, under **Run | Deployment**, select **Deploy to current context (minikube)**, as shown in the following screenshot:

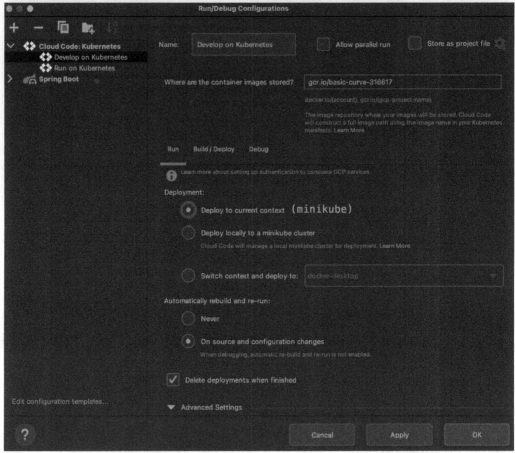

Figure 7.4 – Setting Kubernetes's current context to Minikube

9. Click on **Apply** and **OK** to save your changes.

10. Finally, to run the application on the local Minikube cluster, click on the green run icon:

Figure 7.5 – Running the application

As explained earlier, Cloud Code uses Skaffold. You should see the following output once the application has been deployed successfully to the local Minikube cluster:

```
[INFO] ------------------------------------------------------------------------
[INFO] BUILD SUCCESS
[INFO] ------------------------------------------------------------------------
[INFO] Total time:  20.359 s
[INFO] Finished at: 2021-06-27T02:47:53+05:30
[INFO] ------------------------------------------------------------------------
Starting test...
Tags used in deployment:
 - java-guestbook-backend -> gcr.io/basic-curve-316617/java-guestbook-backend:4fef579ce175876586ed9b0c52c61687a52589dd91f0b72200d698a304f239ee
 - java-guestbook-frontend -> gcr.io/basic-curve-316617/java-guestbook-frontend:fb9a313c20b4db355512687026ea471da57a9ed7b1e9feb2444a9792b798a70b
Starting deploy...
 - service/java-guestbook-backend created
 - service/java-guestbook-frontend created
 - service/java-guestbook-mongodb created
 - deployment.apps/java-guestbook-backend created
 - deployment.apps/java-guestbook-frontend created
 - deployment.apps/java-guestbook-mongodb created
Waiting for deployments to stabilize...
 - deployment/java-guestbook-backend: waiting for init container init-db-ready to start
   - pod/java-guestbook-backend-5c54bb47d6-4m5bt: waiting for init container init-db-ready to start
 - deployment/java-guestbook-frontend: waiting for rollout to finish: 0 of 1 updated replicas are available...
 - deployment/java-guestbook-mongodb: creating container mongo
   - pod/java-guestbook-mongodb-f9d7cc7cd-qfx6m: creating container mongo
 - deployment/java-guestbook-frontend is ready. [2/3 deployment(s) still pending]
 - deployment/java-guestbook-mongodb is ready. [1/3 deployment(s) still pending]
 - deployment/java-guestbook-backend is ready.
Deployments stabilized in 1 minute 9.487 seconds
Press Ctrl+C to exit
Watching for changes...
```

Figure 7.6 – Deployment logs

11. You will receive a notification in the **Event Logs** section in IntelliJ. Click on **View** to access the local URLs of your deployed Kubernetes services:

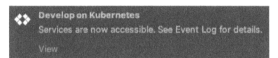

Figure 7.7 – Event logs notification

12. You can click on the **java-guestbook-frontend** URL to access the application:

Figure 7.8 – Available services

You should see the following screen after accessing the `http://localhost:4503` URL:

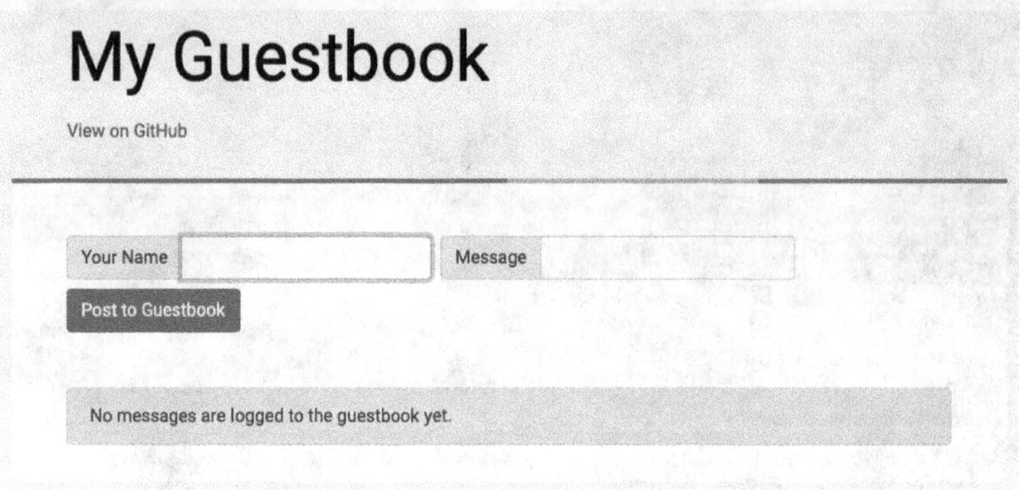

My Guestbook

View on GitHub

Your Name [   ]   Message [   ]

Post to Guestbook

No messages are logged to the guestbook yet.

Figure 7.9 – My Guestbook application landing page

In this section, we installed the Cloud Code plugin and used the startup template provided to start with this plugin quickly. With our very minimal setup, we built and deployed a Java application to the local Kubernetes cluster. The following section will create a Spring Boot application that will display real-time air quality data.

Creating a Spring Boot application

According to WHO (`https://www.who.int/health-topics/air-pollution`), air pollution is killing approximately 7 million people worldwide every year. This is a cause of concern, not only for developed nations but developing nations as well. We should do everything we can to stop this from happening by taking strong measures. We, as technologists, can create solutions to make people aware about the air quality in their area. With this, people can take preventive measures such as wearing masks while they are out and keeping the elderly and kids at home if the air outside is toxic.

In this section, we will create a Spring Boot application that will show real-time air quality data for your current location. We will use the API provided by Openaq (`https://openaq.org/`), a non-profit organization called Wikipedia of air quality data. It exposes many endpoints for real-time air quality data, but we will use the `https://api.openaq.org/v1/latest?country=IN` URL for our Spring Boot application. Let's begin.

As always, we will start by downloading a working stub for our Spring Boot application using Spring Initializr by browsing `https://start.spring.io/`. We will also add the following dependencies for our project:

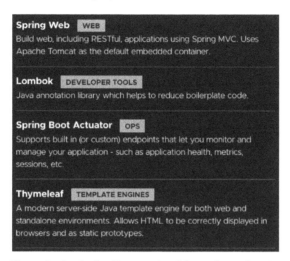

Figure 7.10 – Spring Boot project Maven dependencies

Apart from the dependencies that we've already discussed, we will also add the following Dekorate Spring Boot starter dependency:

```
<dependency>
    <groupId>io.dekorate</groupId>
    <artifactId>kubernetes-spring-starter</artifactId>
    <version>2.1.4</version>
</dependency>
```

Dekorate (`https://github.com/dekorateio/dekorate`) is a tool that generates Kubernetes manifests for you automatically. It can detect if the application has a Spring Boot web dependency and automatically generates Kubernetes manifests during compilation and, by default, configures services, deployments, and probes. Optionally, in your main class, you can add the `@KubernetesApplication` annotation to add some customization. For example, you can provide the number of replicas, service type, ingress, and many other things:

```
@KubernetesApplication(serviceType = ServiceType.LoadBalancer,
replicas = 2,expose = true)
```

Dekorate generates Kubernetes manifests in .json or .yml format in the target/classes/META-INF/dekorate directory.

The following is the code for the Kubernetes service manifests:

```
apiVersion: v1
kind: Service
metadata:
  labels:
    app.kubernetes.io/name: scanner
    app.kubernetes.io/version: 0.0.1-SNAPSHOT
  name: scanner
spec:
  ports:
    - name: http
      port: 8080
      targetPort: 8080
  selector:
    app.kubernetes.io/name: scanner
    app.kubernetes.io/version: 0.0.1-SNAPSHOT
  type: LoadBalancer
```

The following is the relevant part of the deployment Kubernetes manifest. As you can see, Dekorate has generated both liveness and readiness probes:

```
spec:
  containers:
    - env:
        - name: KUBERNETES_NAMESPACE
          valueFrom:
            fieldRef:
              fieldPath: metadata.namespace
      image: breathe
      imagePullPolicy: IfNotPresent
```

```
    livenessProbe:
      failureThreshold: 3
      httpGet:
        path: /actuator/health/liveness
        port: 8080
        scheme: HTTP
      initialDelaySeconds: 0
      periodSeconds: 30
      successThreshold: 1
      timeoutSeconds: 10
    name: scanner
    ports:
      - containerPort: 8080
        name: http
        protocol: TCP
    readinessProbe:
      failureThreshold: 3
      httpGet:
        path: /actuator/health/readiness
        port: 8080
        scheme: HTTP
      initialDelaySeconds: 0
      periodSeconds: 30
      successThreshold: 1
      timeoutSeconds: 10
```

This is the AirQualityController class, which has been annotated with the @Controller annotation. All the incoming HTTP requests to /index are handled by the index() method, which takes the country code, limit, page, and city name as input. The default values for these parameters are IN, 5, 1, and Delhi, respectively.

As per the following code snippet, we have a method named `getAqiForCountry()`, that is called every time we request `/index`. This method also uses `RestTemplate` to fetch real-time air quality data from the endpoint, as mentioned in the `COUNTRY_AQI_END_POINT` variable, and returns an `AqiCountryResponse` object. Refer to the following code:

```java
@Controller
@RequestMapping("/index")
public class AirQualityController {

    private static final String COUNTRY_AQI_END_POINT = "https://api.openaq.org/v1/latest?country";
    private static final Logger LOGGER = LoggerFactory.getLogger(AirQualityController.class);
    private RestTemplate restTemplate;

    @Bean
    private RestTemplate createRestTemplate(RestTemplateBuilder builder) {
        restTemplate = builder.setReadTimeout(Duration.ofSeconds(60)).build();
        return restTemplate;
    }

    private AqiCountryResponse getAqiForCountry(String code, String limit, String page, String city) {
        LOGGER.info("URL " + COUNTRY_AQI_END_POINT + "=" + code + "&" + "limit" + "=" + limit + "&" +
"page" + "=" + page + "&" + "city" + "=" + city);
        return restTemplate.getForObject(COUNTRY_AQI_END_POINT + "=" + code + "&" + "limit" + "=" +
limit + "&" + "page" + "=" + page + "&" + "city" + "=" + city, AqiCountryResponse.class);
    }

    @GetMapping
    public String index(@RequestParam(value = "country", required = true, defaultValue = "IN") String
country,@RequestParam(value = "limit", required = true, defaultValue = "5") String
limit,@RequestParam(value = "page", required = true, defaultValue = "1") String
page,@RequestParam(value = "city", required = true, defaultValue = "Delhi") String city,Model model) {
        var aqiForCountry = getAqiForCountry(country, limit, page, city).getResults();
        for (Location location : aqiForCountry) {
            Collections.sort(location.getMeasurements(),
 Comparator.comparing(Measurement::getParameter).thenComparing(Measurement::getValue));
        }
        model.addAttribute("response", aqiForCountry);
        return "index";
    }
}
```

Figure 7.11 – Code for real-time air quality data

> **Tip**
>
> The `RestTemplate` class has been put in maintenance mode since version 5.0. This means that only minor bug fixes will be allowed and that it will be removed in the future in favor of the `org.springframework.web.reactive.client.WebClient` class, which supports both synchronous and asynchronous operations. To use `WebClient`, you will have to add another dependency, such as `spring-boot-starter-webflux`. If you want to avoid having just one dependency, you can also use the new HTTP Client API, which was added in Java 11. With this new API, we can send requests either synchronously or asynchronously. In the following synchronous blocking example, we are using the `send(HttpRequest, HttpResponse.BodyHandler)` method. This method blocks until the request is sent and a response is received:
>
> ```
> HttpClient httpClient = HttpClient.newBuilder().
> build();
>
> HttpRequest httpRequest = HttpRequest.newBuilder()
>         .uri(URI.create("URL"))
>         .GET()
>         .build();
>
> HttpResponse<String> syncHttpResponse =
> httpClient.send(httpRequest, HttpResponse.
> BodyHandlers.ofString());
> ```
>
> Similarly, for asynchronous non-blocking, we can use the `sendAsync(HttpRequest, HttpResponse.BodyHandler)` method. It returns with a `CompletableFuture<HttpResponse>` that can be combined with different asynchronous tasks.

The `AqiCountryResponse` object contains the following data elements:

```
@Data
@JsonIgnoreProperties(ignoreUnknown = true)
  public class AqiCountryResponse {
    public List<Location> results;
}
@Data
@JsonIgnoreProperties(ignoreUnknown = true)
  public class Location {
    public String location;
```

```
    public String city;
    public List<Measurement> measurements;
}
@Data
@JsonIgnoreProperties(ignoreUnknown = true)
  public class Measurement {
    public String parameter;
    public String value;
    public String unit;
}
```

Finally, we must do some sorting and return the data to the index.html page to render it on the UI. For the UI part, we have used the Spring Boot Thymeleaf dependency. Using the following logic, we can display the real-time air quality data inside a table on the /index.html page:

```html
<div th:if="${ not#lists.isEmpty(response) }">
    <table class="table table-bordered table-striped"
      id="example" style="width: -moz-max-content">
        <tr>
            <th>Location</th>
            <th>City</th>
            <th colspan="30">Measurements</th>
        </tr>
        <tr th:each="response : ${response}">
            <td th:text="${response.location}"></td>
            <td th:text="${response.city}"></td>
            <th:block th:each="p ${response.measurements}">
                <td th:text="${p.parameter}"></td>
                <td th:text="${p.value}+' '+${p.unit}"></td>
            </th:block>
        </tr>
        <table>
</div>
```

We have also created a static HTML table that specifies the air pollution levels, with colors assigned to them inside the same page. These colors make it easy for people to identify if pollution has reached an alarming level or not in their respective areas:

```
<table class="table table-bordered" id="example1"
  style="width: max-content">
    <tr>
        <th>AQI</th>
        <th>Air Pollution Level</th>
        <th>Health Implications</th>
        <th>Cautionary Statement (for PM2.5)</th>
    </tr>
    <tr bgcolor="green">
        <td>0-50</td>
        <td>Good</td>
        <td>Air quality is considered satisfactory,
            and air pollution poses little or no risk</td>
        <td>None</td>
    </tr>
    <tr bgcolor="yellow">
        <td>51-100</td>
        <td>Moderate</td>
        <td>Air quality is acceptable; however,
            for some pollutants there may be a moderate
            health concern for a very small number of
            people who are unusually sensitive to air
            pollution.
        </td>
        <td>Active children and adults, and people with
            respiratory disease, such as asthma,
            should limit prolonged outdoor exertion.
        </td>
    </tr>
<table>
```

At this point, the application is ready. We can try it out by running it using the `mvn sprinboot:run` command. Let's do that and see if we get the expected output. In the following screenshot, you can see that we have changed the default city to Mumbai and that we can view real-time air quality data for Mumbai:

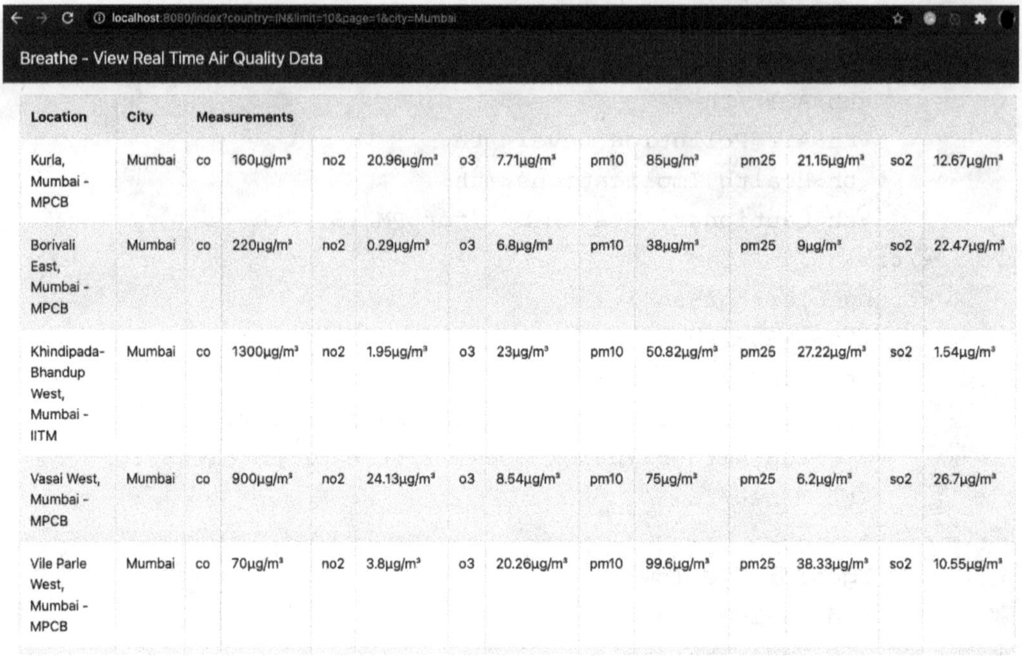

Location	City	Measurements													
Kurla, Mumbai - MPCB	Mumbai	co	160µg/m³	no2	20.96µg/m³	o3	7.71µg/m³	pm10	85µg/m³	pm25	21.15µg/m³	so2	12.67µg/m³
Borivali East, Mumbai - MPCB	Mumbai	co	220µg/m³	no2	0.29µg/m³	o3	6.8µg/m³	pm10	38µg/m³	pm25	9µg/m³	so2	22.47µg/m³
Khindipada-Bhandup West, Mumbai - IITM	Mumbai	co	1300µg/m³	no2	1.95µg/m³	o3	23µg/m³	pm10	50.82µg/m³	pm25	27.22µg/m³	so2	1.54µg/m³
Vasai West, Mumbai - MPCB	Mumbai	co	900µg/m³	no2	24.13µg/m³	o3	8.54µg/m³	pm10	75µg/m³	pm25	6.2µg/m³	so2	26.7µg/m³
Vile Parle West, Mumbai - MPCB	Mumbai	co	70µg/m³	no2	3.8µg/m³	o3	20.26µg/m³	pm10	99.6µg/m³	pm25	38.33µg/m³	so2	10.55µg/m³

Figure 7.12 – Breathe – Real Time Air Quality Data for Mumbai

On the same page, we can see a table that contains information related to different AQI ranges and their severity:

AQI	Air Pollution Level	Health Implications	Cautionary Statement (for PM2.5)
0-50	Good	Air quality is considered satisfactory, and air pollution poses little or no risk	None
51-100	Moderate	Air quality is acceptable; however, for some pollutants there may be a moderate health concern for a very small number of people who are unusually sensitive to air pollution.	Active children and adults, and people with respiratory disease, such as asthma, should limit prolonged outdoor exertion.
101-150	Unhealthy for Sensitive Groups	Members of sensitive groups may experience health effects. The general public is not likely to be affected.	Active children and adults, and people with respiratory disease, such as asthma, should limit prolonged outdoor exertion.
151-200	Unhealthy	Everyone may begin to experience health effects; members of sensitive groups may experience more serious health effects.	Active children and adults, and people with respiratory disease, such as asthma, should avoid prolonged outdoor exertion; everyone else, especially children, should limit prolonged outdoor exertion.
201-300	Very Unhealthy	Health warnings of emergency conditions. The entire population is more likely to be affected.	Active children and adults, and people with respiratory disease, such as asthma, should avoid prolonged outdoor exertion; everyone else, especially children, should limit prolonged outdoor exertion.
300+	Hazardous	Health alert: everyone may experience more serious health effects.	Everyone should avoid all outdoor exertion.

Figure 7.13 – Air quality index

In this section, we created a Spring Boot application that displays the real-time air quality data of a city in your country.

In the next section, we will use the Cloud Code plugin to containerize and deploy our application to our local Kubernetes cluster.

Containerizing and deploying a Spring Boot application using Cloud Code

Let's try to containerize and deploy the Spring Boot application we created in the previous section. To containerize our Spring Boot application, we will use jib-maven-plugin. We've used this many times in previous chapters, so I will skip the setup for it here. We will deploy to a local Minikube cluster using kubectl. Let's learn how to do this:

1. First, we will need a skaffold.yaml file in the root directory of our project.

2. You can create an empty file named skaffold.yaml and use the Cloud Code auto-completion feature, as shown in the following screenshot, to generate a working skaffold.yaml file:

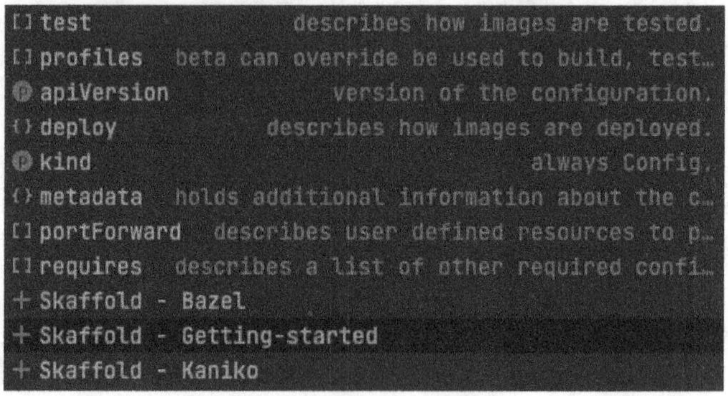

Figure 7.14 – Creating the skaffold.yaml file using Cloud Code

3. Sometimes, a new schema version may be available. Cloud Code is smart enough to detect those changes and will suggest that you upgrade the schema as well, as shown in the following screenshot:

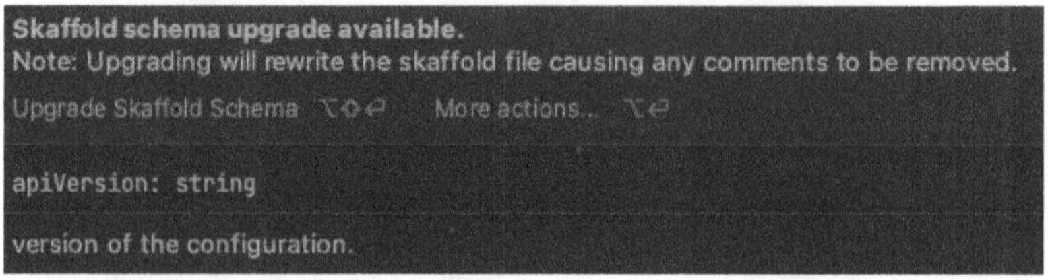

Figure 7.15 – Updating the schema version using Cloud Code

4. The following is the final version of our `skaffold.yaml` configuration file. Here, you can see that we have used `jib` to containerize our application. We used `kubectl` for deployment, and the path we've used is the same as the one we used for Dekorate for our Kubernetes manifest generation:

```yaml
apiVersion: skaffold/v2beta20
kind: Config
metadata:
  name: scanner
build:
  artifacts:
  - image: breathe
    jib:
      project: com.air.quality:scanner
deploy:
  kubectl:
    manifests:
      - target/classes/META-INF/dekorate/kubernetes.yml
```

Soon after creating the `skaffold.yaml` configuration file, Cloud Code detects the change and suggests that we **Create Cloud Code Kubernetes Run Configurations**, as follows:

Figure 7.16 – Creating run configurations using Cloud Code

5. Upon clicking this, under the **Run/Debug** configuration in IntelliJ, two new profiles will be created called **Develop on Kubernetes** and **Run on Kubernetes**:

Figure 7.17 – Cloud Code profiles

6. To run our application in continuous development mode, choose **Develop on Kubernetes** from the dropdown. Cloud Code internally uses the `skaffold dev` command in this mode. It will do the following for you:

    ▪ It will start watching for changes in your source code.

    ▪ It will containerize our Spring Boot application using Jib. Since we are using a local Kubernetes cluster, Skaffold is smart enough not to push the image to the remote registry for a fast inner development loop. Instead, it will load the image to the local Docker daemon.

    ▪ It will deploy the image to the Minikube cluster, port forward to port `8080`, and start streaming the logs in your IDE. The events logs in your IDE will show the service URL, which you can use to access your application. The output will be similar to what we saw in the previous section.

The **Run on Kubernetes** option is similar to the `skaffold run` command. You can use this option to deploy when you want instead of doing so on every code change.

Even though we have not done that, you can even use Cloud Code to deploy to a remote Kubernetes cluster. If your Kubernetes context is pointed toward a remote cluster such as GKE, then Cloud Code can do the deployment there as well. If you don't have a remote cluster, then Cloud Code can also help you create that.

Cloud Code has a good integration for running serverless workloads, as well as using Google's Cloud Run.

In this section, you learned how to containerize and deploy a Spring Boot application to a local Kubernetes cluster using Cloud Code. Now, let's summarize this chapter.

Summary

In this chapter, you learned how to use the Cloud Code plugin developed by Google to do a single-click deployment of your Kubernetes applications from your IDE. We started this chapter by explaining the various features of Cloud Code. In the example, we explained how we could use the startup templates provided by Cloud Code to write, build, and deploy your Java applications from your IDE. Then, we created a Spring Boot application that uses Dekorate to generate Kubernetes manifests at compile time. Finally, we containerized and deployed the Spring Boot application to a local Minikube cluster.

By doing this, you have discovered how you can use Cloud Code to increase your productivity while developing cloud-native applications.

The next chapter will talk about how we can deploy a Spring Boot application to the Google Kubernetes Engine.

8

Deploying a Spring Boot Application to the Google Kubernetes Engine Using Skaffold

In the previous chapter, you learned how to deploy a **Spring Boot** application to a local **Kubernetes** cluster using **Google**'s **Cloud Code** plugin for **IntelliJ**. This chapter focuses on deploying the same Spring Boot application to the remote **Google Kubernetes Engine** (**GKE**), a managed Kubernetes service provided by the **Google Cloud Platform** (**GCP**). We will introduce you to Google's recently launched serverless Kubernetes offering, **GKE Autopilot**. You will also get to know **Google Cloud SDK** and **Cloud Shell**, and use them to connect and manage a remote Kubernetes cluster.

In this chapter, we're going to cover the following main topics:

- Getting started with the Google Cloud Platform
- Working with Google Cloud SDK and Cloud Shell
- Setting up the Google Kubernetes Engine
- Introducing GKE Autopilot clusters
- Deploying a Spring Boot application to the GKE

By the end of this chapter, you will have a solid understanding of the essential services provided by the GCP to deploy a Spring Boot application to Kubernetes.

Technical requirements

You'll need the following to be installed on your system to follow the examples in this chapter:

- Eclipse (`https://www.eclipse.org/downloads/`) or IntelliJ IDE (`https://www.jetbrains.com/idea/download/`)
- Git (`https://git-scm.com/downloads`)
- Google Cloud SDK
- GCP account
- Spring Boot 2.5
- OpenJDK 16

The code examples present in the chapter can also be found on GitHub at `https://github.com/PacktPublishing/Effortless-Cloud-Native-App-Development-Using-Skaffold`.

Getting started with the Google Cloud Platform

Today, many organizations take advantage of services provided by different cloud providers such as **Amazon Web Services (AWS)**, Google's GCP, **Microsoft Azure**, **IBM Cloud**, or **Oracle Cloud**. The advantage of using these cloud vendors is that you don't have to manage infrastructure yourself, and you typically pay per hour for the use of these servers. Also, most of the time, if the organizations are unaware of or fail to address the computing power needed for their applications, it might result in the overprovision of computing resources.

If you're managing the infrastructure yourself, you have to keep an army of people to take care of upkeep activities such as patching an operating system, upgrading the software, and upgrading the hardware. These cloud vendors help us solve business problems by providing these services for us. Also, you get built-in maintenance for the products that these cloud vendors support, whether it is databases or managed services such as Kubernetes. If you have already used any of these cloud vendors, you might find that all of these vendors provide similar services or products. Still, the implementation and how they work are different.

For example, you can see the services provided by GCP and their AWS and Azure equivalents in the link `https://cloud.google.com/free/docs/aws-azure-gcp-service-comparison`.

Now we know that there are advantages of using these cloud vendors for different use cases, let's talk about one such cloud vendor – Google Cloud Platform.

Google Cloud Platform, often abbreviated to GCP, provides you with a collection of services such as on-demand virtual machines (through **Google Compute Engine**), object storage for storing files (through **Google Cloud Storage**), and managed Kubernetes (through Google Kubernetes Engine), to name a few.

Before you can begin utilizing Google's Cloud services, you are first required to sign up for an account. If you already have a Google account such as a **Gmail** account, then you can use that to log in, but you are still required to sign up separately for the Cloud account. You can skip this step if you are already signed up on the Google Cloud Platform.

First, navigate to `https://cloud.google.com`. Next, you will be asked to go through a typical Google sign-in process. If you don't have a Google account yet, follow the sign-up process to create one. The following screenshot is the Google Cloud sign-in page:

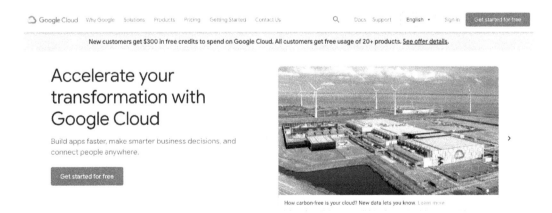

Figure 8.1 – Getting started with Google Cloud

If you look at the screenshot carefully, it says **New customers get $300 in free credits to spend on Google Cloud. All customers get free usage of 20+ products**. This means you can use free tier products without paying anything, and you will also get credit worth $300 for 90 days to explore or evaluate different services provided by GCP. For example, you can use Compute Engine, Cloud Storage, and **BigQuery** free of charge within specified monthly usage limits.

You can either click on **Get started for free** or **Sign In**. You must provide your billing information if you sign up for the first time, and this redirects you to your Cloud **Console**. Also, a new project is automatically created for you. A project is a sort of workspace for your work. All the resources in a single project are isolated from those in all the other projects. You can control access to this project and only grant access to specific individuals or service accounts. The following screenshot is the view of your Google Cloud Console dashboard:

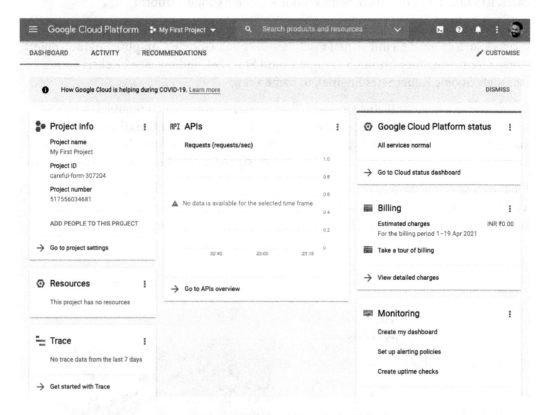

Figure 8.2 – Google Cloud Console dashboard

On the left side of the Console page, you can view different services offered by GCP:

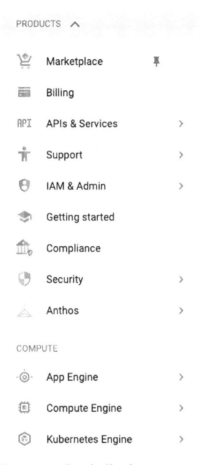

Figure 8.3 – Google Cloud services view

In this chapter, the focus will be on the GKE services API provided by GCP. But before we discuss these, we need to install some tools to use those services. Let's talk about those tools in the next section.

Working with Google Cloud SDK and Cloud Shell

You have access to the GCP Console now, and you can pretty much do anything using the Console. But a better approach for a developer is to use Cloud SDK, which is a collection of tools that allow faster local development by using emulators or tools like **kubectl**, **Skaffold**, and **minikube**. Not only that, but you can manage your resources, authenticate with remote Kubernetes clusters, and enable or disable GCP services from your local workstation. Another option is to use Cloud Shell from your browser, and we will be exploring both options in this chapter. Cloud SDK gives you tools and a library for interacting with its product and services. You can install and remove components as per your needs when using Cloud SDK.

Let's start with Cloud SDK. You can navigate to `https://cloud.google.com/sdk/` and click on the **Get Started** button. This will redirect you to the installation guide. A minimum prerequisite for Cloud SDK is to have Python. Supported versions are Python 3 (3.5 to 3.8 preferred) and Python 2 (2.7.9 or higher). For example, modern versions of macOS include the appropriate version of Python required for Cloud SDK. However, if you'd like to install Python 3 with Cloud SDK, you can choose the macOS 64-bit with bundled Python installation offering.

Downloading Cloud SDK on Linux

Cloud SDK requires Python to be installed, so first verify the Python version using the following command:

```
python --version
```

To download the Linux 64-bit archive file from your command line, run the following command:

```
curl -O https://dl.google.com/dl/cloudsdk/channels/rapid/
downloads/google-cloud-sdk-336.0.0-linux-x86_64.tar.gz
```

For the 32-bit archive file, run the following command:

```
curl -O https://dl.google.com/dl/cloudsdk/channels/rapid/
downloads/google-cloud-sdk-336.0.0-linux-x86.tar.gz
```

Downloading Cloud SDK on macOS

To download Cloud SDK on macOS, you have the following options to choose from:

Platform	Package	Size	SHA256 Checksum
macOS 64-bit (x86_64)	google-cloud-sdk-336.0.0-darwin-x86_64.tar.gz	85.0 MB	e1f049c536491e77ff9bfb8e29755ba7006a43c7dd8911350848b668025f6039
macOS 64-bit (arm64)	google-cloud-sdk-336.0.0-darwin-arm.tar.gz	84.9 MB	9d6dfd1d09a813a7e9a99304d7f76b980d5526482165fc7f3b5cb4002b3fdf39
macOS 64-bit with bundled Python (x86_64)	google-cloud-sdk-336.0.0-darwin-x86_64-bundled-python.tar.gz	126.9 MB	65255e20c4ecd68ff89899c35e4db62ec18fcb6a6e9ccc72b912c6a13f634133
macOS 32-bit (x86)	google-cloud-sdk-336.0.0-darwin-x86.tar.gz	88.8 MB	ab6974c6f627e4bec9474169d15165c9ddacc5f8b1b3076b47e07ec29e84d77b

Figure 8.4 – Download options for macOS

If you are not sure about your machine's hardware, then run the uname –m command. Based upon your machine, you will get the following output:

```
$uname -m
x86_64
```

Now select the appropriate package and download it from the URL given in the **Package** column in the table available at `https://cloud.google.com/sdk/docs/install#mac`.

Setting up Cloud SDK

After downloading the package, you need to extract the archive to a location of your choice on your filesystem. The following is the content of the `google-cloud-sdk` extracted archive:

```
tree -L 1 google-cloud-sdk
google-cloud-sdk
├── LICENSE
├── README
├── RELEASE_NOTES
├── VERSION
├── bin
├── completion.bash.inc
├── completion.zsh.inc
├── data
├── deb
├── install.bat
├── install.sh
├── lib
├── path.bash.inc
├── path.fish.inc
├── path.zsh.inc
├── platform
├── properties
└── rpm
```

After extracting the archive, you can proceed with the installation by running the `install.sh` script available in the root directory of your extract. You might see the following output:

```
$ ./google-cloud-sdk/install.sh
Welcome to the Google Cloud SDK!
To help improve the quality of this product, we collect
anonymized usage data
and anonymized stacktraces when crashes are encountered;
additional information
```

```
is available at <https://cloud.google.com/sdk/usage-
statistics>. This data is

handled in accordance with our privacy policy
<https://cloud.google.com/terms/cloud-privacy-notice>. You may
choose to opt in this

collection now (by choosing 'Y' at the below prompt), or at any
time in the

future by running the following command:

    gcloud config set disable_usage_reporting false


Do you want to help improve the Google Cloud SDK (y/N)?  N
Your current Cloud SDK version is: 336.0.0
The latest available version is: 336.0.0
```

In the following screen, you can see the list of installed and not installed components:

```
                                    Components
┌───────────────┬──────────────────────────────────────────────────────┬──────────────────────────┬──────────┐
│    Status     │                        Name                          │            ID            │   Size   │
├───────────────┼──────────────────────────────────────────────────────┼──────────────────────────┼──────────┤
│ Not Installed │ App Engine Go Extensions                             │ app-engine-go            │  4.8 MiB │
│ Not Installed │ Appctl                                               │ appctl                   │ 18.5 MiB │
│ Not Installed │ Cloud Bigtable Command Line Tool                     │ cbt                      │  7.6 MiB │
│ Not Installed │ Cloud Bigtable Emulator                              │ bigtable                 │  6.6 MiB │
│ Not Installed │ Cloud Datalab Command Line Tool                      │ datalab                  │  < 1 MiB │
│ Not Installed │ Cloud Datastore Emulator                             │ cloud-datastore-emulator │ 18.4 MiB │
│ Not Installed │ Cloud Firestore Emulator                             │ cloud-firestore-emulator │ 41.9 MiB │
│ Not Installed │ Cloud Pub/Sub Emulator                               │ pubsub-emulator          │ 60.4 MiB │
│ Not Installed │ Cloud SQL Proxy                                      │ cloud_sql_proxy          │  7.4 MiB │
│ Not Installed │ Emulator Reverse Proxy                               │ emulator-reverse-proxy   │ 14.5 MiB │
│ Not Installed │ Google Cloud Build Local Builder                     │ cloud-build-local        │  6.2 MiB │
│ Not Installed │ Google Container Registry's Docker credential helper │ docker-credential-gcr    │  2.2 MiB │
│ Not Installed │ Kustomize                                            │ kustomize                │ 22.8 MiB │
│ Not Installed │ Minikube                                             │ minikube                 │ 23.7 MiB │
│ Not Installed │ Nomos CLI                                            │ nomos                    │ 22.5 MiB │
│ Not Installed │ On-Demand Scanning API extraction helper             │ local-extract            │ 11.5 MiB │
│ Not Installed │ Skaffold                                             │ skaffold                 │ 17.5 MiB │
│ Not Installed │ anthos-auth                                          │ anthos-auth              │ 16.3 MiB │
│ Not Installed │ config-connector                                     │ config-connector         │ 44.9 MiB │
│ Not Installed │ gcloud Alpha Commands                                │ alpha                    │  < 1 MiB │
│ Not Installed │ gcloud Beta Commands                                 │ beta                     │  < 1 MiB │
│ Not Installed │ gcloud app Java Extensions                           │ app-engine-java          │ 53.1 MiB │
│ Not Installed │ gcloud app PHP Extensions                            │ app-engine-php           │ 21.9 MiB │
│ Not Installed │ gcloud app Python Extensions                         │ app-engine-python        │  6.1 MiB │
│ Not Installed │ gcloud app Python Extensions (Extra Libraries)       │ app-engine-python-extras │ 27.1 MiB │
│ Not Installed │ kpt                                                  │ kpt                      │ 12.2 MiB │
│ Not Installed │ kubectl                                              │ kubectl                  │  < 1 MiB │
│ Not Installed │ kubectl-oidc                                         │ kubectl-oidc             │ 16.3 MiB │
│ Not Installed │ pkg                                                  │ pkg                      │          │
│ Installed     │ BigQuery Command Line Tool                           │ bq                       │  < 1 MiB │
│ Installed     │ Cloud SDK Core Libraries                             │ core                     │ 17.9 MiB │
│ Installed     │ Cloud Storage Command Line Tool                      │ gsutil                   │  3.9 MiB │
└───────────────┴──────────────────────────────────────────────────────┴──────────────────────────┴──────────┘
```

Figure 8.5 – List of Google Cloud SDK components

You can use the following Cloud SDK commands to install or remove components:

```
To install or remove components at your current SDK version
[336.0.0], run:
  $ gcloud components install COMPONENT_ID
  $ gcloud components remove COMPONENT_ID
Enter a path to an rc file to update, or leave blank to use
[/Users/ashish/.zshrc]:
No changes necessary for [/Users/ashish/.zshrc].
For more information on how to get started, please visit:
  https://cloud.google.com/sdk/docs/quickstarts
```

Make sure you source your bash profile after this using the source .zshrc command. From the installation, you can see that only three components, . bq, core, and gsutil, are installed by default.

The next step is to run gcloud init to initialize the SDK using the following command:

```
$/google-cloud-sdk/bin/gcloud init
Welcome! This command will take you through the configuration
of gcloud.
Your current configuration has been set to: [default]
You can skip diagnostics next time by using the following flag:
  gcloud init --skip-diagnostics
Network diagnostic detects and fixes local network connection
issues.
Checking network connection...done
Reachability Check passed.
Network diagnostic passed (1/1 checks passed).
You must log in to continue. Would you like to log in (Y/n)?  Y
Your browser has been opened to visit:
```

```
    https://accounts.google.com/o/oauth2/auth?re-
sponse_type=code&client_id=32555940559.apps.goog-
leusercontent.com&redirect_uri=http%3A%2F%2Flocal-
host%3A8085%2F&scope=openid+https%3A%2F%2Fwww.googleapis.
com%2Fauth%2Fuserinfo.email+https%3A%2F%2Flocal-
host%3A8085%2F&scope=openid+https%3A%2F%2Fwww.googleapis.
com%2Fauth%2Fuserinfo.email+https%3A%2F%2Fwww.googleapis.
com%2Fauth%2Fcloud-platform+https%3A%2F%2Fwww.googleapis.
com%2Fauth%2Fappengine.admin+https%3A%2F%2Fwww.googleapis.
com%2Fauth%2Fcompute+https%3A%2F%2Fwww.googleapis.com%2Fau-
th%2Faccounts.reauth&state=CU1Yhij0NWZB8kZvNx6aAslkkXdlY-
f&access_type=offline&code_challenge=sJO_hf6-zNKLjVSw9fZlx-
jLodFA-EsunnBWiRB5snmw&code_challenge_method=S256
```

At this point, you will be redirected to a browser window and will be asked to log in to your Google account for authentication and granting access to Cloud SDK for your Cloud resources.

On clicking the **Allow** button, it will make sure that next time you can interact with the GCP API as yourself. After granting access, you will see the following screen confirming the authentication:

Figure 8.6 – Google Cloud SDK authentication completed

Now you have authenticated yourself and are ready to work with Cloud SDK. You may see the following on the command line after completing the authentication:

```
Updates are available for some Cloud SDK components.  To
install them,
please run:
  $ gcloud components update


You are logged in as: [XXXXXXX@gmail.com].


Pick cloud project to use:
 [1] xxxx-xxx-307204
 [2] Create a new project
Please enter numeric choice or text value (must exactly match
list
item): 1
Your current project has been set to: [your-project-id].
Do you want to configure a default Compute Region and Zone?
(Y/n)? Y
Which Google Compute Engine zone would you like to use as
project
default?
If you do not specify a zone via a command line flag while
working
with Compute Engine resources, the default is assumed.
 [1] us-east1-b
 [2] us-east1-c
 [3] us-east1-d
 ................
Please enter a value between 1 and 77, or a value present in
the list:  1
Your project default Compute Engine zone has been set to
[us-east1-b].
You can change it by running [gcloud config set compute/zone
NAME].
Your project default Compute Engine region has been set to
[us-east1].
You can change it by running [gcloud config set compute/region
NAME].
```

```
Your Google Cloud SDK is configured and ready to use!
. . . . . .
```

From the command line output, it is clear we have selected the project and confirmed the Compute Engine region. Now, we have successfully installed Cloud SDK. In the next section, we will learn about Cloud Shell.

Using Cloud Shell

Cloud Shell is a browser-based terminal/CLI and editor. It comes pre-installed with tools like Skaffold, minikube, and Docker, to name a few. It can be activated by clicking the following icon, available at the top right-hand side of the Cloud Console browser window:

Figure 8.7 – Activating Cloud Shell

You will be redirected to the following screen after activation:

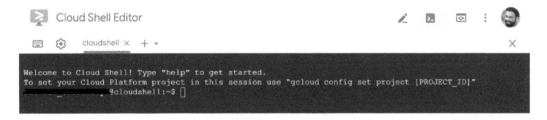

Figure 8.8 – Cloud Shell editor

You can set your project ID by using the `gcloud config set project projectid` command, or just start playing around with `gcloud` commands. The following are some of the highlighted features that Cloud Shell provides:

- Cloud Shell is entirely browser-based, and you can access it from anywhere. The only requirement is internet connectivity.

- Cloud Shell gives you a 5 GB persistent storage mounted to your `$HOME` directory.

- Cloud Shell comes with an online code editor. You can use it to build, test, and debug your applications.

- Cloud Shell also comes with a Git client installed so that you can clone and push changes to your repository from the code editor or command line.

- Cloud Shell comes with a web preview in which you can view your local changes in a web app.

We have installed and configured Google Cloud SDK for our use. We have also looked at Cloud Shell and the features it provides. Now let's create a Kubernetes cluster where we can deploy our Spring Boot application.

Setting up a Google Kubernetes Engine cluster

We would need to set up a Kubernetes cluster on GCP to deploy our containerized Spring Boot application. GCP can provide a hosted and managed deployment of Kubernetes. We can create a Kubernetes cluster on GCP using the following two methods:

- Creating a Kubernetes cluster using Google Cloud SDK

- Creating a Kubernetes cluster using Google Console

Let's discuss each of these in detail.

Creating a Kubernetes cluster using Google Cloud SDK

We can create a Kubernetes cluster for running containers using the following gcloud SDK command. This will create a Kubernetes cluster with default settings:

```
ashish@MacBook-Air ~ % gcloud container clusters create sample-cluster

WARNING: Starting in January 2021, clusters will use the Regular release channel by default when `--cluster-version`, `--release-c
hannel`, `--no-enable-autoupgrade`, and `--no-enable-autorepair` flags are not specified.
WARNING: Currently VPC-native is not the default mode during cluster creation. In the future, this will become the default mode an
d can be disabled using `--no-enable-ip-alias` flag. Use `--[no-]enable-ip-alias` flag to suppress this warning.
WARNING: Starting with version 1.18, clusters will have shielded GKE nodes by default.
WARNING: Your Pod address range (`--cluster-ipv4-cidr`) can accommodate at most 1008 node(s).
WARNING: Starting with version 1.19, newly created clusters and node-pools will have COS_CONTAINERD as the default node image when
 no image type is specified.
Creating cluster sample-cluster in us-east1-b...done.
Created [https://container.googleapis.com/v1/projects/basic-curve-316617/zones/us-east1-b/clusters/sample-cluster].
To inspect the contents of your cluster, go to: https://console.cloud.google.com/kubernetes/workload_/gcloud/us-east1-b/sample-clu
ster?project=basic-curve-316617
kubeconfig entry generated for sample-cluster.
NAME            LOCATION     MASTER_VERSION    MASTER_IP      MACHINE_TYPE   NODE_VERSION      NUM_NODES   STATUS
sample-cluster  us-east1-b   1.20.8-gke.2100   34.75.135.86   e2-medium      1.20.8-gke.2100   3           RUNNING
```

Figure 8.9 – GKE cluster up and running

We have successfully created a Kubernetes cluster using Cloud SDK. Next, we will try to create the cluster using Google Console.

Creating a Kubernetes cluster using Google Console

To create a Kubernetes cluster using the Console, you should first use the left-hand side navigation bar and choose **Kubernetes Engine**. In the presented option, select **Clusters**:

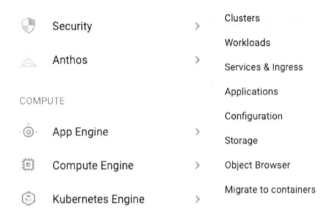

Figure 8.10 – Getting started with Google Kubernetes Engine cluster creation

After that, you will see the following screen on the next page:

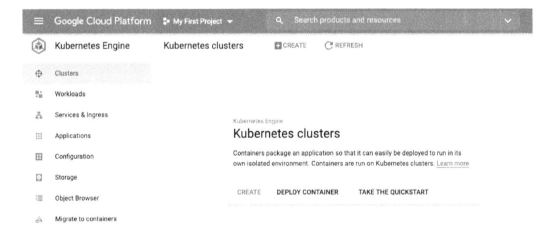

Figure 8.11 – Creating a Google Kubernetes Engine cluster

You can choose to create the cluster by clicking on the **CREATE** button on the popup, or by clicking **+CREATE** at the top of the page. Both will give you the following options to choose from, as explained in *Figure 8.12*:

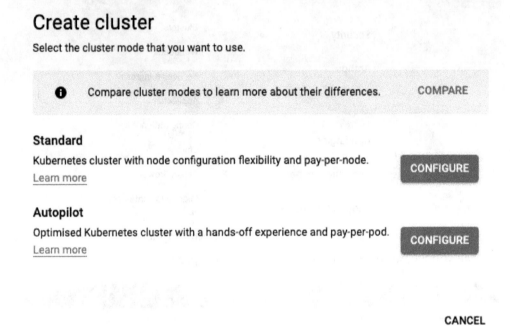

Figure 8.12 – Google Kubernetes Engine cluster modes

You can choose to create a **Standard** Kubernetes cluster or a completely hands-off experience with **Autopilot** mode. In this section, we will discuss the Standard cluster. We will cover the Autopilot separately in the next section.

In the Standard cluster mode, you have the flexibility to choose the number of nodes for your cluster and to tweak the configurations or setup as per your needs. The following is a walkthrough of creating a Kubernetes cluster. Since we are going with default configurations, you must click **Next** to accept the default options.

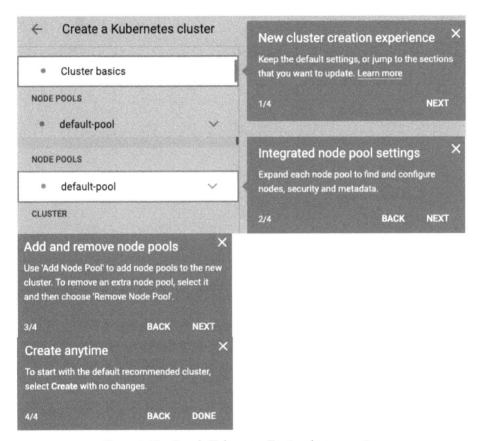

Figure 8.13 – Google Kubernetes Engine cluster creation

Finally, click on the **Create** button at the bottom of the page and voila, your Kubernetes cluster will be up and running in a few minutes!

The following are the default configurations for your cluster:

Figure 8.14 – Google Kubernetes Engine cluster configuration view

Your Kubernetes cluster is now up and running. In the following screenshot, we can see that we have a three-node cluster with six vCPUs and 12 GB of total memory:

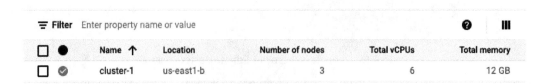

Figure 8.15 – Google Kubernetes Engine cluster up and running

You can view more details about your cluster nodes, storage, and view logs by clicking on the cluster name **cluster-1**. The following are the details of the cluster nodes we just created:

← Clusters   ✏ EDIT   🗑 DELETE   ➕ ADD NODE POOL   ➕ DEPLOY   📲 CONNECT   🗐 DUPLICATE   ⚙ OPERATIONS

Node pools

⇆ Filter   Filter node pools   ❓ ▥

Name ↑	Status	Version	Number of nodes	Machine type	Image type	Auto-scaling	
default-pool	✓ OK	1.18.16-gke.502	3	e2-medium	Container-optimised OS with Docker (cos)	Off	🗑

Nodes

⇆ Filter   Filter nodes   ❓ ▥

Name ↑	Status	CPU requested	CPU allocatable	Memory requested	Memory allocatable	Storage requested	Storage allocatable
gke-cluster-1-default-pool-9de1a9de-20q3	✓ Ready	463 mCPU	940 mCPU	377.49 MB	2.96 GB	0 B	0 B
gke-cluster-1-default-pool-9de1a9de-dq5b	✓ Ready	463 mCPU	940 mCPU	377.49 MB	2.96 GB	0 B	0 B
gke-cluster-1-default-pool-9de1a9de-sn1n	✓ Ready	379 mCPU	940 mCPU	615.51 MB	2.96 GB	0 B	0 B

Figure 8.16 – Google Kubernetes Engine cluster view

You can see that overall cluster status and node health is OK. The cluster nodes are created using Compute Engine GCP, and offering to have machine type as **e2-medium**. You can verify this by viewing the Compute Engine resources on the left-hand side navigation bar. We have the same three nodes shown here, and the GKE cluster uses these that we have just created.

🖳 Compute Engine   VM instances   🗎 CREATE INSTANCE   ⬆ IMPORT VM   ↻ REFRESH   ▶ START/RESUME

Virtual machines ⌃

🖳 VM instances

🖳 Instance templates

🖳 Sole-tenant nodes

🖳 Machine images

🖳 TPUs

🖳 Migrate for Compute Engine

🖳 Committed use discounts

	INSTANCES	INSTANCE SCHEDULE

⇆ Filter   Enter property name or value

☐	●	Name ↑	Zone	Recommendations	In use by
☐	✓	gke-cluster-1-default-pool-9de1a9de-20q3	us-east1-b		gke-cluster-1-default-pool-9de1a9de-grp
☐	✓	gke-cluster-1-default-pool-9de1a9de-dq5b	us-east1-b		gke-cluster-1-default-pool-9de1a9de-grp
☐	✓	gke-cluster-1-default-pool-9de1a9de-sn1n	us-east1-b		gke-cluster-1-default-pool-9de1a9de-grp

Figure 8.17 – Google Kubernetes Engine cluster VM instances

We have learned how to create a Kubernetes Standard cluster using Google Console. In the next section, we will learn about the Autopilot cluster.

Introducing the Google Kubernetes Engine Autopilot cluster

On February 24th, 2021, Google announced the general availability of their fully managed Kubernetes services, GKE Autopilot. It is a completely managed and serverless Kubernetes as a service offering. No other cloud provider currently offers this level of automation when managing the Kubernetes cluster on the cloud. Most cloud providers leave some cluster management for you, be it managing the control planes (**API server**, **etcd**, **scheduler**, and so on), worker nodes, or creating everything from scratch as per your needs.

GKE Autopilot, as the name suggests, is an entirely hands-off experience, and in most cases you only have to specify a cluster name and region, set the network if you want to, and that's it. You can focus on deploying your workloads and let Google fully manage your Kubernetes cluster. Google is offering 99.9% uptime for Autopilot pods in multiple zones. Even if you manage this yourself, you will not beat the number that Google is offering. On top of this, GKE Autopilot is cost-effective as you don't pay for **Virtual Machines** (**VMs**), and you are only billed per second for resources (for example, vCPU, memory, and disk space consumed by your pods).

So what's the difference between a GKE Standard cluster like the one we created in the previous section and a GKE Autopilot cluster? The answer is as follows: with the Standard cluster, you manage only the nodes, as the GKE manages the control plane, and with GKE Autopilot, you don't manage anything (not even your worker nodes).

This raises a question: is it a good or a bad thing that I cannot control my nodes? Now, this is debatable, but most organizations today are not handling traffic or loads like amazon. com, google.com, or netflix.com. It may be an oversimplification, but to be honest, even if you think you have specific needs or you need a specialized cluster, more often than not, you end up wasting a lot of time and resources in securing and managing your cluster. If you have a team of SRE that can match the level of experience or knowledge of Google SRE, you can do whatever you like with your cluster. But most organizations today don't have such expertise and don't know what they are doing. That's why it is better to rely on fully managed Kubernetes services such as GKE Autopilot – it is battle-tested and hardened based on the best practices learned from Google SRE.

We have talked enough about GKE Autopilot features and the complete abstraction it provides over how we manage the Kubernetes cluster. However, keeping these abstractions in mind, there are some restrictions as well. For example, you cannot run privileged mode for containers in the Autopilot mode. For a complete list of limitations, read the official documentation at `https://cloud.google.com/kubernetes-engine/docs/concepts/autopilot-overview#limits`.

We have gained enough knowledge about GKE Autopilot so far, and now it's time to create our cluster. So, let's begin!

Creating an Autopilot cluster

After clicking on the **Configure** button, as explained in *Figure 8.13*, you will be redirected to the following screen:

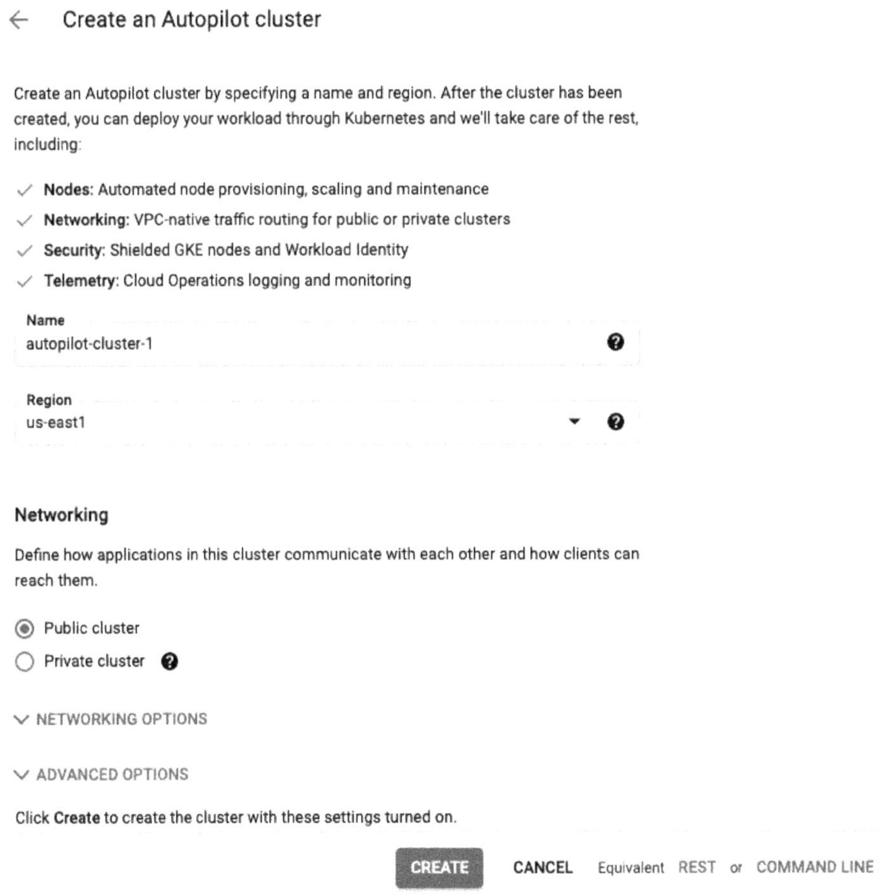

Figure 8.18 – Creating a GKE Autopilot cluster

Autopilot clusters have features such as node management, networking, security, and telemetry already built-in with Google-recommended best practices. GKE Autopilot makes sure that your cluster is optimized and production-ready.

As you can see, you have very few options to change here. You can change the **Name** of the cluster, pick another **Region**, or choose the **Networking** (that is, public or private). Under **NETWORKING OPTIONS**, you can change things like network, subnetwork, pod IP address range and cluster services IP address range, and so on, as seen in the following screenshot:

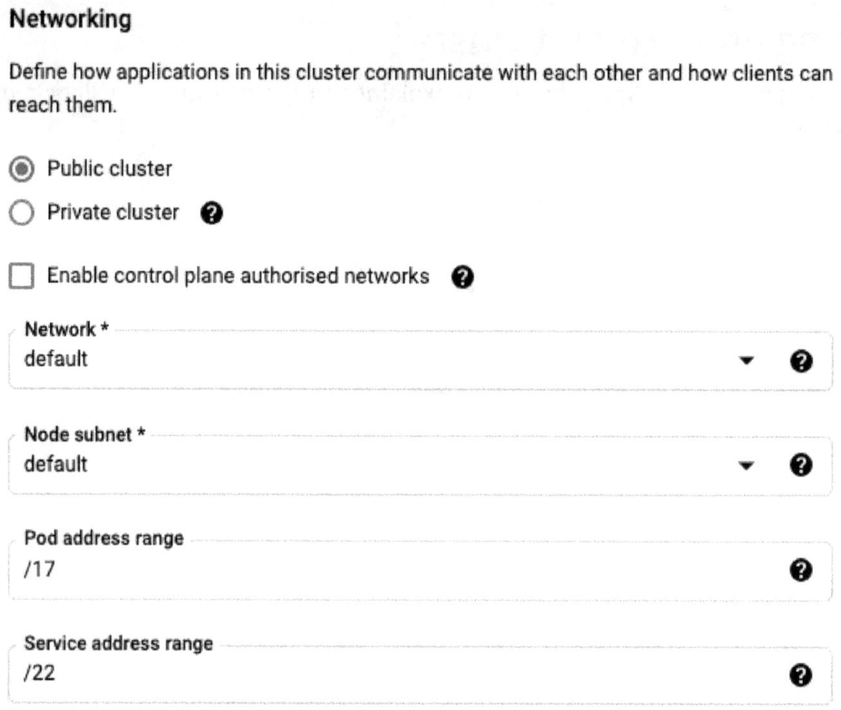

Networking

Define how applications in this cluster communicate with each other and how clients can reach them.

- ◉ Public cluster
- ○ Private cluster ❓

- ☐ Enable control plane authorised networks ❓

Network *
default   ▼   ❓

Node subnet *
default   ▼   ❓

Pod address range
/17   ❓

Service address range
/22   ❓

Figure 8.19 – GKE Autopilot cluster configurations

Under **ADVANCED OPTIONS**, you can enable a maintenance window and also allow maintenance exclusion for a specific time range (as seen in *Figure 8.19*). In this window, your GKE cluster will go for an automated maintenance window and will not be available for your use. You should choose a maintenance window as per your needs.

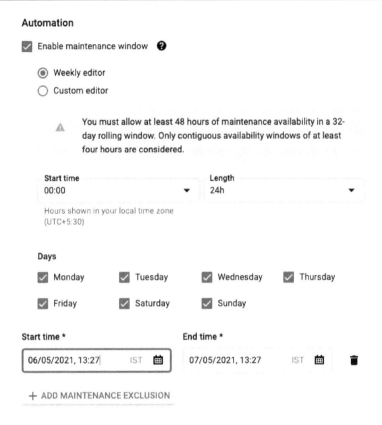

Figure 8.20 – Configuring a GKE Autopilot cluster maintenance window

For now, we will go with default values and click on the **CREATE** button at the bottom of the page to create the cluster. It may take a few minutes to create your cluster. In the following screenshot, you can see the Autopilot cluster is up and running:

Figure 8.21 – GKE Autopilot cluster up and running

Here, you can see that the number of nodes is not mentioned, as it is managed by GKE.

Next, we can try connecting to this cluster. To do so, click on the three dots at the top right of your screen and click on **Connect**:

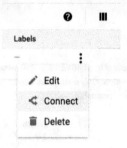

Figure 8.22 – Connecting to a GKE Autopilot cluster

After clicking on **Connect**, the following popup should appear. You can copy the command mentioned here into your CLI or Cloud Shell:

Connect to the cluster

You can connect to your cluster via command-line or using a dashboard.

Command-line access

Configure kubectl ⧉ command line access by running the following command:

```
$ gcloud container clusters get-credentials autopilot-cluster-1 --region us-east1 --project careful-form-307204
```

RUN IN CLOUD SHELL

Cloud Console dashboard

You can view the workloads running in your cluster in the Cloud Console Workloads dashboard .

OPEN WORKLOADS DASHBOARD

OK

Figure 8.23 – Commands for connecting to a GKE Autopilot cluster

Then you can verify the cluster details by using the following `kubectl get nodes` command:

```
$ kubectl get nodes

NAME                                                      STATUS   ROLES    AGE     VERSION
gk3-autopilot-cluster-1-default-pool-333bdb98-ngbl        Ready    <none>   3h54m   v1.18.16-
gke.2100

gk3-autopilot-cluster-1-default-pool-9d750cdd-gk17        Ready    <none>   3h55m   v1.18.16-
gke.2100

$ kubectl get nodes

gk3-autopilot-cluster-1-default-pool-333bdb98-ngbl        Ready    <none>   3h54m   v1.18.16-
gke.2100

gk3-autopilot-cluster-1-default-pool-9d750cdd-gk17        Ready    <none>   3h55m   v1.18.16-
gke.2100
```

Figure 8.24 – kubectl command output

We can also create GKE cluster in autopilot mode using the following command:

```
ashish@MacBook-Air ~ % gcloud container clusters create-auto gke-autopilot-cluster1 --region=us-east1

WARNING: Starting with version 1.18, clusters will have shielded GKE nodes by default.
WARNING: The Pod address range limits the maximum size of the cluster. Please refer to https://cloud.google.com/kubernetes-engine/
docs/how-to/flexible-pod-cidr to learn how to optimize IP address allocation.
WARNING: Starting with version 1.19, newly created clusters and node-pools will have COS_CONTAINERD as the default node image when
 no image type is specified.
Creating cluster gke-autopilot-cluster1 in us-east1...done.
Created [https://container.googleapis.com/v1/projects/basic-curve-316617/zones/us-east1/clusters/gke-autopilot-cluster1].
To inspect the contents of your cluster, go to: https://console.cloud.google.com/kubernetes/workload_/gcloud/us-east1/gke-autopilo
t-cluster1?project=basic-curve-316617
kubeconfig entry generated for gke-autopilot-cluster1.
NAME                     LOCATION    MASTER_VERSION    MASTER_IP       MACHINE_TYPE    NODE_VERSION      NUM_NODES   STATUS
gke-autopilot-cluster1   us-east1    1.20.9-gke.701    35.227.108.35   e2-medium       1.20.9-gke.701    3           RUNNING
```

Figure 8.25 – GKE cluster in autopilot

We can further verify this on Google Cloud Console as well. You can see that we now have two clusters. The first is created using Cloud Console and the second using the command line with gcloud.

		Name ↑	Location	Mode	Number of nodes	Total vCPUs	Total memory	Notifications	Labels	
☐	●	autopilot-cluster-1	us-east1	Autopilot		0	0 GB		—	⋮
☐	✓	gke-autopilot-cluster1	us-east1	Autopilot		0	0 GB		—	⋮

Figure 8.26 – GKE Autopilot clusters

We have gone through the different ways of creating the Kubernetes cluster on GCP. Now, let's deploy a working Spring Boot application to the GKE using Skaffold.

Deploying a Spring Boot application to the GKE

The Spring Boot application that we will use in this section is the same as in the previous chapter (the application we named *Breathe – View Real-Time Air Quality Data*). We are already familiar with the application, so we will directly jump to the deployment to the GKE. We will be using `gke-autopilot-cluster1` we created in the previous section for deployment. We will do the deployment using the following two approaches using Skaffold:

- Deploying from local to a remote GKE cluster using Skaffold
- Deploying from Cloud Shell to a GKE cluster using Skaffold

Deploying from local to a remote GKE cluster using Skaffold

In this section, you will learn how you can deploy the Spring Boot application to a remote Kubernetes cluster with the help of Skaffold. Let's begin:

1. In the previous chapter, we used **Dockerfile** to containerize our Spring Boot application. However, in this chapter, we will be using the `Jib-Maven` plugin to containerize the application. We already know how to use the jib-maven plugin from previous chapters, so we will skip explaining this again here.

2. The only change is that we will be using the **Google Container Registry** (**GCR**) for storing the image pushed by Jib. GCR is a secure private registry for your images. Before that, we would need to make sure that GCR access is enabled for your account. You can allow access by using the following `gcloud` command:

```
gcloud services enable containerregistry.googleapis.com
```

Or you can navigate to `https://cloud.google.com/container-registry/docs/quickstart` and enable the Container Registry API by clicking on the **Enable the API** button.

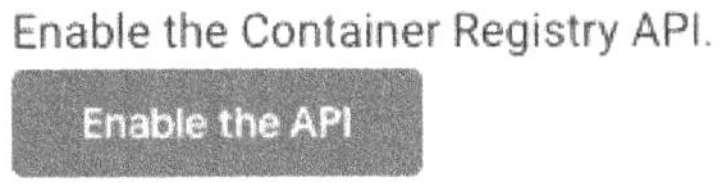

Enable the Container Registry API.

Enable the API

Figure 8.27 – Enabling Google Container Registry API

Next, you will be asked to choose a project and then click on **Continue**. That's it!

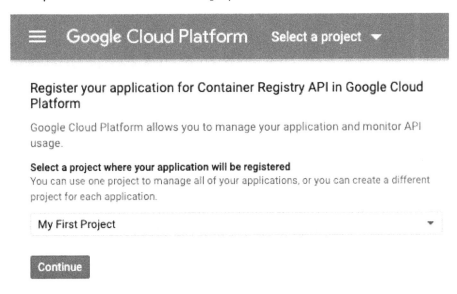

Figure 8.28 – Register your application for Container Registry API

3. You can make the images under your container registry available for public access as well. Users of your images can pull the images without any authentication if they are public. In the following screenshot, you can see an option, **Enable Vulnerability Scanning**, for images pushed to your container registry. If you want, you can allow it to scan your container images for vulnerabilities.

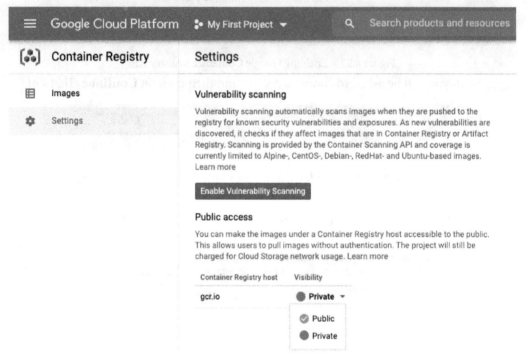

Figure 8.29 – GCR settings

4. The next piece of the puzzle is to create Kubernetes manifests such as **Deployment** and **Service**. In the previous chapter, we created them using the **Dekorate** tool (`https://github.com/dekorateio/dekorate`). We will be using the same Kubernetes manifest generation process here as well. The generated Kubernetes manifests are available under the `target/classes/META-INF/dekorate/kubernetes.yml` path.

5. Next, we will run the `skaffold init --XXenableJibInit` command, which will create a `skaffold.yaml` configuration file for us. You can see that Skaffold added the Kubernetes manifest's path in the generated `deploy` section of the `skaffold.yaml` file and will use `jib` for image building:

```
apiVersion: skaffold/v2beta20
kind: Config
metadata:
  name: scanner
build:
  artifacts:
  - image: breathe
    jib:
      project: com.air.quality:scanner
deploy:
  kubectl:
    manifests:
    - target/classes/META-INF/dekorate/kubernetes.yml
```

6. We have the same main class as explained in the previous chapter, which uses the @KubernetesApplication (serviceType = ServiceType.LoadBalancer) annotation provided by the Dekorate tool to declare the service type as `LoadBalancer`:

```
@KubernetesApplication(serviceType = ServiceType.
LoadBalancer)
@SpringBootApplication
public class AirQualityScannerApplication {

    public static void main(String[] args) {
        SpringApplication.run(AirQualityScannerApplication.
        class, args);
    }

}
```

At the time of compilation, Dekorate will generate the following Kubernetes manifests. I have also kept them in the k8s directory in the source code, as sometimes we have to manually add or remove things from Kubernetes manifests. The Deployment and Service Kubernetes manifests can also be found on GitHub at `https://github.com/PacktPublishing/Effortless-Cloud-Native-App-Development-Using-Skaffold/blob/main/Chapter07/k8s/kubernetes.yml`.

After that, we need to make sure that you are authenticated to use Google Cloud services using the `gcloud auth list` command. You will see the following output:

```
Credentialed AccountsACTIVE   ACCOUNT*          <my_
account>@<my_domain.com>To set the active account,
run:     $ gcloud config set account 'ACCOUNT'
```

If you are not authenticated, you can also use the `gcloud auth login` command.

7. If it is not already set, set your GCP project using the `gcloud config set project <PROJECT_ID>` command.

8. Make sure that the Kubernetes context is set to remote Google Kubernetes cluster. Use the following command to verify that:

```
$ kubectl config current-context
gke_project_id_us-east1_gke-autopilot-cluster1
```

9. We are now ready for deployment. Let's run the `skaffold run --default-repo=gcr.io/<PROJECT_ID>` command. This will build the container image of the application. Push it to the remote GCR.

Figure 8.30 – Image pushed to Google Container Registry

The pushed image details can be seen in the following screenshot:

Figure 8.31 – Google Container Registry image view

10. Finally, deploy it to a remote Google Kubernetes cluster. It takes some time to stabilize the deployment when you run it for the first time, but subsequent runs are much faster.

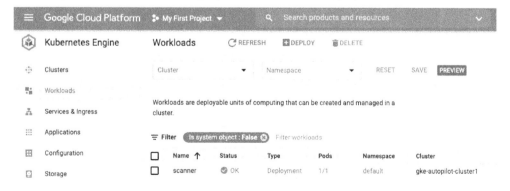

Figure 8.32 – Skaffold run output

11. We can also view the Deployment status on Google Cloud Console. Go to **Kubernetes Engine**, then click on the **Workloads** tab on the left-hand side navigation bar to view the deployment status. The Deployment status is **OK**.

Figure 8.33 – Deployment status

You can view further **Deployment** details by clicking on the application's name.

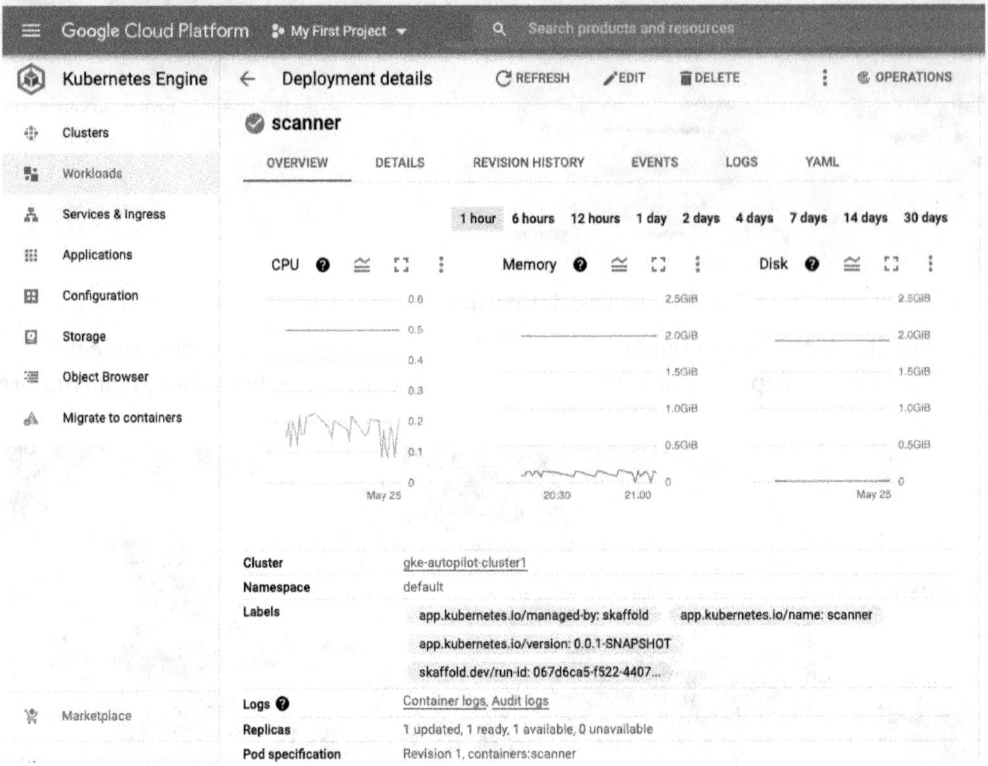

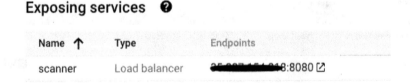

Figure 8.34 – Deployment details

12. Everything looks great so far. Now we just need the IP address of the service so that we can access our application. On the same Deployment details page at the bottom, we have details about our service.

Exposing services ❓

Name ↑	Type	Endpoints
scanner	Load balancer	▓▓▓▓▓▓▓▓:8080 ☐

Figure 8.35 – Exposed services

13. Let's hit the URL and verify if we get the desired output or not. We can view the real-time air quality data for Delhi:

Location	City	Measurements											
NSIT Dwarka, Delhi - CPCB	Delhi	co	260µg/m³	no2	11.74µg/m³	o3	39.24µg/m³	pm10	57.38µg/m³	pm25	33.6µg/m³	so2	5.88µg/m³
DTU, Delhi - CPCB	Delhi	co	360µg/m³	no2	23µg/m³	o3	-2.8µg/m³	pm10	0µg/m³	pm25	1000µg/m³	so2	5.7µg/m³
Shadipur, Delhi - CPCB	Delhi	co	1020µg/m³	no2	11.14µg/m³	o3	45.92µg/m³	pm10	60.79µg/m³	pm25	3.3µg/m³	so2	13.9µg/m³
Punjabi Bagh, Delhi - DPCC	Delhi	co	900µg/m³	no2	7.2µg/m³	o3	69.5µg/m³	pm10	80µg/m³	pm25	61µg/m³	so2	18.5µg/m³
Mandir Marg, Delhi - DPCC	Delhi	co	800µg/m³	no2	52.3µg/m³	o3	32µg/m³	pm10	95µg/m³	pm25	68µg/m³	so2	12.8µg/m³

Figure 8.36 – Spring Boot application response

14. We can verify the health of the application using actuator `/health/liveness` and `/health/readiness` endpoints. We have used these endpoints as liveness and readiness probes for pods deployed to the Kubernetes cluster.

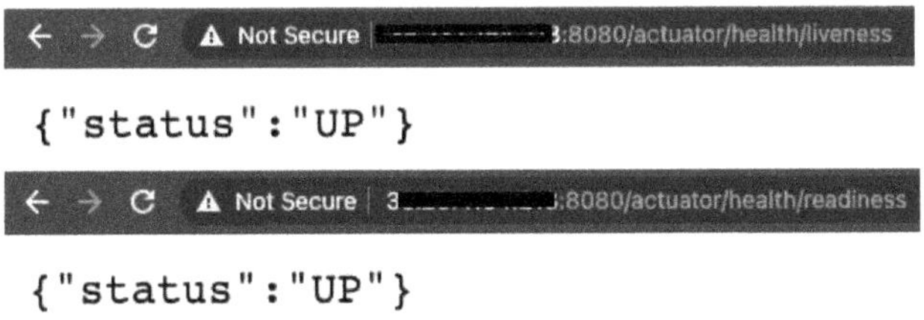

```
{"status":"UP"}
```

```
{"status":"UP"}
```

Figure 8.37 – Spring Boot application actuator probes

With these steps, we have completed the deployment of our Spring Boot application to a remote Google Kubernetes cluster from a local workstation using Skaffold. In the next section, we will learn about deploying the application from a browser-based Cloud Shell environment to the GKE.

Deploying from Cloud Shell to a GKE cluster using Skaffold

In this section, the focus will be on deploying the Spring Boot application to the GKE using the browser-based Cloud Shell tool. Let's begin!

1. The first step is to activate the Cloud Shell environment. This can be done by clicking on the **Activate Cloud Shell** icon in the top-right corner of Google Cloud Console.

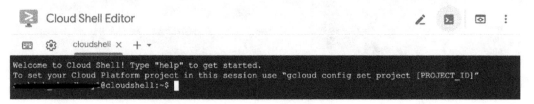

Figure 8.38 – Cloud Shell editor

2. As you can see in the previous screenshot, you are asked to set your Cloud PROJECT_ID with the gcloud config set project [PROJECT_ID] command. You can use this if you know your PROJECT_ID or use commands like gcloud projects list. After this, Cloud Shell would ask for permission to authorize your request by making a call to the GCP API. You don't have to provide credentials for each request after this authorization.

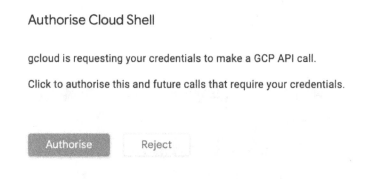

Figure 8.39 – Authorizing Cloud Shell

We need the source code of the application in the Cloud Shell environment. Cloud Shell comes with a Git client installed, so we can run the git clone https://github.com/PacktPublishing/Effortless-Cloud-Native-App-Development-using-Skaffold.git command and clone our GitHub repository.

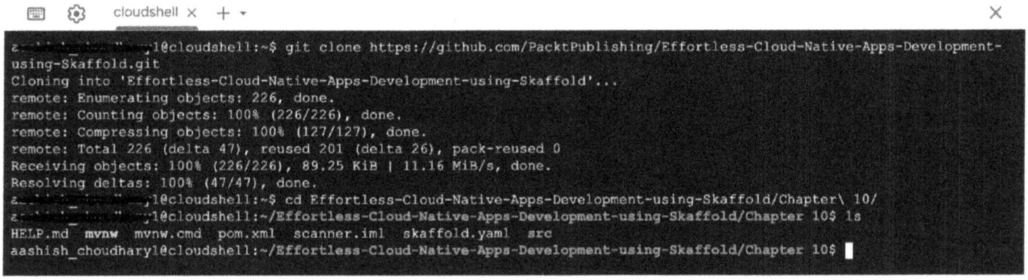

Figure 8.40 – Cloning the GitHub repository

3. Next, you need to compile the project so that you can generate Kubernetes manifests. Run the `./mvnw clean compile` command to build your project. Your build will fail and you will get an error:

    ```
    [ERROR] Failed to execute goal org.apache.maven.
    plugins:maven-compiler-plugin:3.8.1:compile (default-
    compile) on project scanner: Fatal error compiling:
    error: invalid target release: 16 -> [Help 1] .
    ```

    The reason for this failure is that `JAVA_HOME` is set to Java 11 in the Cloud Shell environment:

    ```
    $ java -version
    openjdk version "11.0.11" 2021-04-20
    OpenJDK Runtime Environment (build 11.0.11+9-post-Debian-
    1deb10u1)
    OpenJDK 64-Bit Server VM (build 11.0.11+9-post-Debian-
    1deb10u1, mixed mode, sharing)
    ```

    We have specified in `pomx.ml` to use Java 16. This problem can be solved by downloading Java 16 and setting the `JAVA_HOME` environment variable.

    > **Note**
    >
    > We have the right tool to solve this problem, **SDKMAN**, that can be accessed from `https://sdkman.io/`. It allows you to work in parallel with multiple versions of **Java JDK**s. Check out the supported JDKs (`https://sdkman.io/jdks`) and SDKs (`https://sdkman.io/sdks`). With the new six months release cycle, we get a new JDK every six months. As developers, we love to try and explore these features by manually downloading and changing JAVA_HOME if we need to switch to a different JDK. This whole process is manual, and with `SDKMAN`, we just have run a single command to download a JDK of your choice, and after downloading, it will even update JAVA_HOME to the latest downloaded JDK. Cool, isn't it?

4. Let's try to install JDK16 with SDKMAN. Note that you don't have to install SDKMAN in your Cloud Shell provisioned VM instance as it comes pre-installed. Now enter sdk in your CLI, and it will show you the supported commands:

```
                      @cloudshell:~$ sdk

Usage: sdk <command> [candidate] [version]
       sdk offline <enable|disable>

    commands:
        install    or i     <candidate> [version] [local-path]
        uninstall  or rm    <candidate> <version>
        list       or ls    [candidate]
        use        or u     <candidate> <version>
        completion          <bash|zsh>
        config
        default    or d     <candidate> [version]
        home       or h     <candidate> <version>
        env        or e     [init|install|clear]
        current    or c     [candidate]
        upgrade    or ug    [candidate]
        version    or v
        broadcast  or b
        help
        offline             [enable|disable]
        selfupdate          [force]
        update
        flush               [archives|tmp|broadcast|version]

    candidate  :   the SDK to install: groovy, scala, grails, gradle, kotlin, etc.
                   use list command for comprehensive list of candidates
                   eg: $ sdk list
    version    :   where optional, defaults to latest stable if not provided
                   eg: $ sdk install groovy
    local-path :   optional path to an existing local installation
                   eg: $ sdk install groovy 2.4.13-local /opt/groovy-2.4.13
```

Figure 8.41 – SDKMAN commands help

To learn about different supported JDKs, run the `sdk list java` command. In the following screenshot, you won't be able to see all of the supported JDK vendors, but you get the idea:

```
                    loudshell:~$ sdk list java
================================================================================
Available Java Versions
================================================================================
 Vendor         | Use | Version      | Dist   | Status    | Identifier
--------------------------------------------------------------------------------
 AdoptOpenJDK   |     | 16.0.1.j9    | adpt   |           | 16.0.1.j9-adpt
                |     | 16.0.1.hs    | adpt   |           | 16.0.1.hs-adpt
                |     | 16.0.0.j9    | adpt   |           | 16.0.0.j9-adpt
                |     | 16.0.0.hs    | adpt   |           | 16.0.0.hs-adpt
                |     | 11.0.11.j9   | adpt   |           | 11.0.11.j9-adpt
                |     | 11.0.11.hs   | adpt   |           | 11.0.11.hs-adpt
                |     | 11.0.10.j9   | adpt   |           | 11.0.10.j9-adpt
                |     | 11.0.10.hs   | adpt   |           | 11.0.10.hs-adpt
                |     | 8.0.292.j9   | adpt   |           | 8.0.292.j9-adpt
                |     | 8.0.292.hs   | adpt   |           | 8.0.292.hs-adpt
                |     | 8.0.282.j9   | adpt   |           | 8.0.282.j9-adpt
                |     | 8.0.282.hs   | adpt   |           | 8.0.282.hs-adpt
 Alibaba        |     | 11.0.9.4     | albba  |           | 11.0.9.4-albba
                |     | 8.5.5        | albba  |           | 8.5.5-albba
 Amazon         |     | 16.0.1.9.1   | amzn   |           | 16.0.1.9.1-amzn
                |     | 16.0.0.36.1  | amzn   |           | 16.0.0.36.1-amzn
                |     | 15.0.2.7.1   | amzn   |           | 15.0.2.7.1-amzn
                |     | 11.0.11.9.1  | amzn   |           | 11.0.11.9.1-amzn
                |     | 11.0.10.9.1  | amzn   |           | 11.0.10.9.1-amzn
                |     | 8.292.10.1   | amzn   |           | 8.292.10.1-amzn
                |     | 8.282.08.1   | amzn   |           | 8.282.08.1-amzn
 Azul Zulu      |     | 16.0.1       | zulu   |           | 16.0.1-zulu
                |     | 16.0.0       | zulu   |           | 16.0.0-zulu
                |     | 16.0.0.fx    | zulu   |           | 16.0.0.fx-zulu
                |     | 15.0.2.fx    | zulu   |           | 15.0.2.fx-zulu
                |     | 11.0.11      | zulu   |           | 11.0.11-zulu
                |     | 11.0.10      | zulu   |           | 11.0.10-zulu
                |     | 11.0.10.fx   | zulu   |           | 11.0.10.fx-zulu
```

Figure 8.42 – SDKMAN supported JDKs

To download a vendor-specific JDK, run the `sdk install java Identifier` command. In our case, the actual command will be `sdk install java 16-open`, since we have decided to use the OpenJDK build of Java 16.

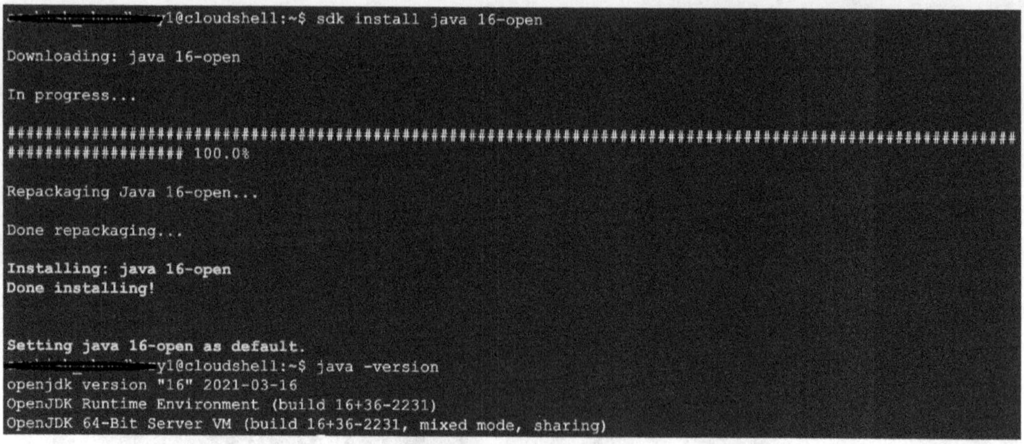

```
y1@cloudshell:~$ sdk install java 16-open

Downloading: java 16-open

In progress...

########################################################################################## 100.0%

Repackaging Java 16-open...

Done repackaging...

Installing: java 16-open
Done installing!

Setting java 16-open as default.
y1@cloudshell:~$ java -version
openjdk version "16" 2021-03-16
OpenJDK Runtime Environment (build 16+36-2231)
OpenJDK 64-Bit Server VM (build 16+36-2231, mixed mode, sharing)
```

Figure 8.43 – Installing JDK16

You might also want to run the following command to change the JDK in your active shell session:

```
$ sdk use java 16-open
Using java version 16-open in this shell.
```

5. Let's compile the project again by running the `./mvnw clean compile` command. In the following output, you can see that the build is successful:

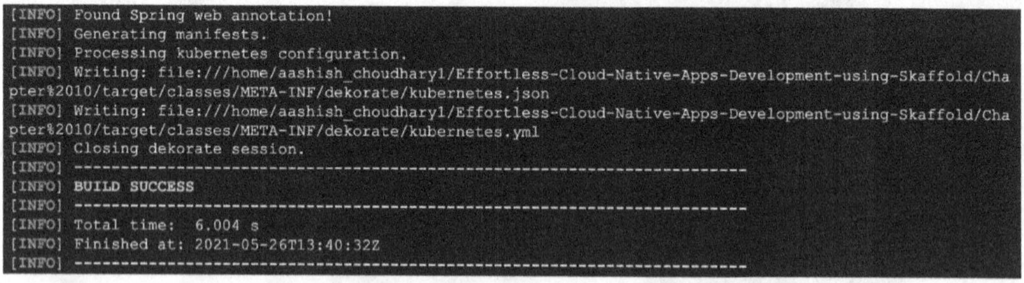

```
[INFO] Found Spring web annotation!
[INFO] Generating manifests.
[INFO] Processing kubernetes configuration.
[INFO] Writing: file:///home/aashish_choudhary1/Effortless-Cloud-Native-Apps-Development-using-Skaffold/Cha
pter%2010/target/classes/META-INF/dekorate/kubernetes.json
[INFO] Writing: file:///home/aashish_choudhary1/Effortless-Cloud-Native-Apps-Development-using-Skaffold/Cha
pter%2010/target/classes/META-INF/dekorate/kubernetes.yml
[INFO] Closing dekorate session.
[INFO] ------------------------------------------------------------------------
[INFO] BUILD SUCCESS
[INFO] ------------------------------------------------------------------------
[INFO] Total time:  6.004 s
[INFO] Finished at: 2021-05-26T13:40:32Z
[INFO] ------------------------------------------------------------------------
```

Figure 8.44 – Maven build success

6. We are ready to run the command to deploy the Spring Boot application to the remote GKE cluster from Cloud Shell. Before that, make sure that your Kubernetes context is set to the remote cluster. If you are not sure, then verify it by running the `kubectl config current-context` command. If it is not set, then set it using the `gcloud container clusters get-credentials gke-autopilot-cluster1 --region us-east1` command, which will add the entry in the `kubeconfig` file.

7. In the last step, we just have to run the `skaffold run --default-repo=gcr.io/<PROJECT_ID>` command. The deployment is stabilized, and the final output is going to be the same as seen in *step 13* in the previous section.

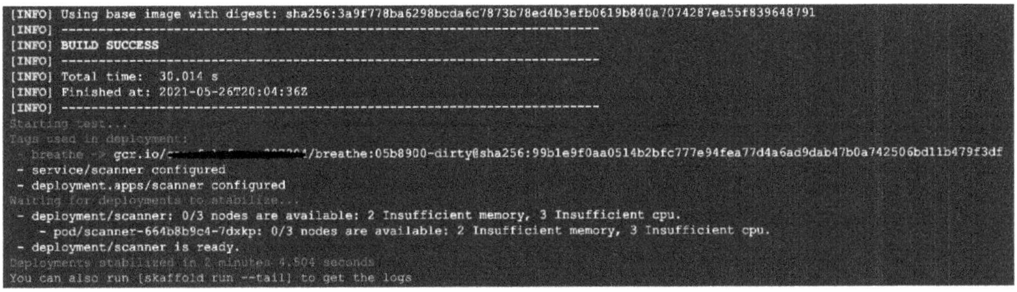

Figure 8.45 – Skaffold run output

This completes the deployment of the Spring Boot application to the remote GKE cluster using the browser-based Cloud Shell environment. We have learned how we can leverage the browser-based preconfigured Cloud Shell environment for development purposes. If you want to play around with and try things, then this is an excellent feature provided by Google. However, I am not sure if you should be using it for your production use cases. The Google Compute Engine VM instances that are provisioned with Cloud Shell are provided on a per-user, per-session basis. Your VM instances will persist if your session is active; otherwise, they will get discarded. For information about the working of Cloud Shell, please go through the official documentation:

`https://cloud.google.com/shell/docs/how-cloud-shell-works`

Summary

In this chapter, we started by discussing the features and advantages of using cloud vendors. Then, we introduced you to the GCP. First, we covered a detailed walkthrough of how you can get onboard to the Cloud Platform. Next, we covered Google Cloud SDK, which allows you to perform various tasks such as installing components, creating Kubernetes clusters, and enabling different services such as Google Container Registry and more.

We also discussed the browser-based Cloud Shell editor, which is powered by Google Compute Engine VM instances. You can use this as a temporary sandbox environment to test various services supported by GCP. Then, we looked at two different ways of creating a Kubernetes cluster using Cloud SDK and Cloud Console. After that, we introduced you to the serverless Kubernetes offering, GKE Autopilot, and covered its features and advantages over standard Kubernetes clusters. Finally, we successfully deployed a Spring Boot application to the GKE Autopilot cluster using Skaffold from local and then Google Cloud Shell in the last section.

In this chapter, you have gained practical knowledge of GCP's managed Kubernetes service, as well as tools like Cloud SDK and Cloud Shell. You have also learned how you can use Skaffold to deploy a Spring Boot application to a remote Kubernetes cluster.

In the next chapter, we will learn about creating a CI/CD pipeline using GitHub actions and Skaffold.

Further reading

- Learn more about GKE autopilot: `https://cloud.google.com/blog/products/containers-kubernetes/introducing-gke-autopilot`

- Learn more about Google Cloud Platform: `https://cloud.google.com/docs`

- Google Cloud Platform for Architects: `https://www.packtpub.com/product/google-cloud-platform-for-architects/9781788834308`

9
Creating a Production-Ready CI/CD Pipeline with Skaffold

In the previous chapter, we learned how to deploy a Spring Boot application to Google Cloud Platform using Skaffold. In this chapter, the focus will be on introducing you to GitHub Actions and their related concepts. We will also demonstrate how we can create a production-ready **continuous integration (CI)** and **continuous deployment (CD)** pipeline of a Spring Boot application using Skaffold and GitHub Actions. In the last section, we will get familiarized with GitOps concepts and learn about creating a continuous delivery pipeline for Kubernetes applications using Argo CD and Skaffold.

In this chapter, we're going to cover the following main topics:

- Getting started with GitHub Actions
- Creating a GitHub Action workflow
- Creating a CI/CD pipeline with GitHub Actions and Skaffold
- Implementing a GitOps workflow with Argo CD and Skaffold

By the end of this chapter, you will have a solid understanding of how you can create an effective CI/CD pipeline using GitHub Actions and Skaffold.

Technical requirements

- Eclipse (https://www.eclipse.org/downloads/) or IntelliJ IDE (https://www.jetbrains.com/idea/download/)
- GitHub account
- Spring Boot 2.5
- OpenJDK 16

The code from the GitHub repository can be found at https://github.com/PacktPublishing/Effortless-Cloud-Native-App-Development-using-Skaffold/tree/main/Chapter07.

Getting started with GitHub Actions

GitHub Actions allows you to build, test, and deploy your workloads from your GitHub repository GitHub Actions is event-driven; for example, when someone creates a pull request, opens an issue, does a deployment, and so on. The specific actions are triggered based upon the events. You can even create your own GitHub Actions to customize the workflow based upon your use case. There is a great marketplace available too, at https://github.com/marketplace, from where you can integrate existing GitHub Actions into your workflow.

GitHub Actions uses a YAML syntax file to define events, jobs, actions, and commands. In the following diagram, you can see a complete list of GitHub Actions components:

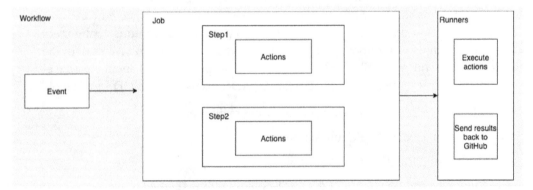

Figure 9.1 – GitHub Actions components

Let's discuss the GitHub components in detail:

- **Workflow**: This is used to build, test, package, release, or deploy the project on GitHub. A workflow consists of jobs and is triggered by events. The workflow is defined in a YAML syntax file available in your GitHub repository inside the `.github/workflows` directory.

- **Events**: This represents an activity that triggers a workflow; for example, pushing changes to a branch or creating a pull request.

- **Jobs**: This consists of steps that are executed on a runner. It uses steps to control the order in which actions are performed. You can run multiple jobs for your workflow. They can be run in parallel or sequentially.

- **Steps**: These represent an action, that is, checking out source code or shell command.

- **Actions**: These represent commands that you would like to run, such as checking out your source code or downloading JDK.

- **Runners**: This is a server hosted on GitHub that has the runner application installed. You can host your own runner or use the one provided by GitHub. Your jobs defined in the workflow are executed on the runner machine. It sends the results, progress, and logs back to the GitHub repository. GitHub-hosted runners support Ubuntu Linux, Microsoft Windows, and macOS.

Now we have learned details about GitHub Action components. In the next section, we will create a GitHub Action workflow for a Spring Boot application.

Creating a GitHub Actions workflow

In this section, we will create a workflow that will build a Spring Boot Java application with GitHub Actions. This workflow will build a Spring Boot application using the `mvn clean install` Maven build tool command. The following is an example of a workflow file of building a Java project with Maven:

```
name: Build Java project with Maven
on:
  push:
    branches: [ main ]
  pull_request:
    branches: [ main ]
jobs:
```

```
build:
  runs-on: ubuntu-latest
  steps:
  - uses: actions/checkout@v2
  - name: Install and Setup Java 16
    uses: AdoptOpenJDK/install-jdk@v1
    with:
      version: '16'
      architecture: x64
  - name: Build with Maven
    run: mvn clean install
```

Here is the explanation of the workflow YAML file:

1. In the workflow YAML file, we have subscribed to a push and pull request event. So whenever a pull request is raised or change is pushed for the main branch, this workflow will trigger.

2. Then inside the jobs section, first we have specified that the job will run on a ubuntu Linux operating system runner hosted by GitHub.

3. In steps, we have defined the actions that need to be executed for this workflow.

4. First, we are checking out the source code on the runner using actions/ checkout@v2.

5. Then we are installing dependencies such as Java. We doing that using the AdoptOpenJDK/install-jdk@v1 action.

6. In the last and final step, we are building a Java project using the mvn clean install command.

Let's see this workflow in action. So next, we will create this workflow in our GitHub repository and trigger the workflow by pushing changes to the main branch.

We will be using the Spring Boot application we created in *Chapter 7, Building and Deploying a Spring Boot Application with the Cloud Code Plugin,* for this demonstration. I have already explained the application in detail so I will not be explaining it again here:

1. The first step is to create the workflow YAML file inside your GitHub repository. This can be done by navigating to the **Actions** tab in your GitHub repository and clicking on the **set up a workflow yourself** link as shown in the following screenshot:

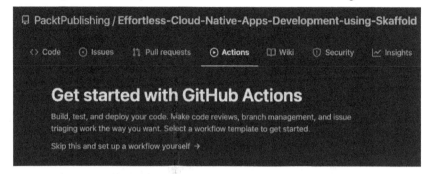

Figure 9.2 – Getting started with GitHub Actions

In the next screen, paste the content of the workflow YAML file that we discussed earlier. Refer to the following screenshot:

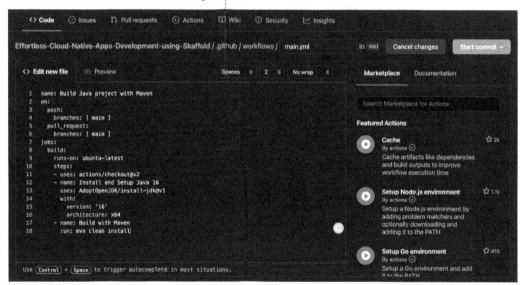

Figure 9.3 – Creating a workflow

After you click on the **Start commit** button, a new commit message window will open where you can enter the commit message.

2. Then click on **Commit new file** to add the workflow file to the GitHub repository:

Figure 9.4 – Committing the workflow file

3. Inside the repository, now you can see that there is a `.github/workflows` directory and inside that directory, we have the `main.yml` workflow file:

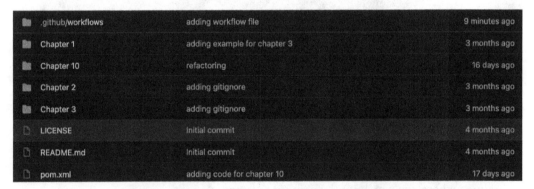

Figure 9.5 – GitHub workflow file added to your repository

This also creates a commit and pushes a change to the repository, which triggers the workflow. In the following screenshot, you can see the workflow is triggered and in progress:

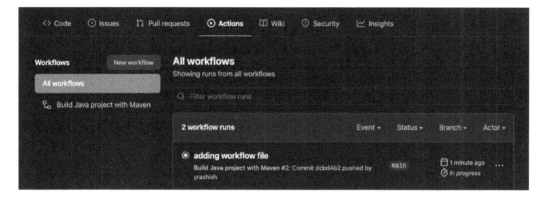

Figure 9.6 – Executing the GitHub workflow

In the following screenshot, you can see the pipeline is green and the triggered workflow has completed successfully:

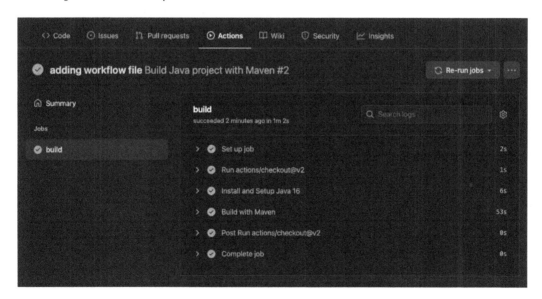

Figure 9.7 – GitHub workflow completed successfully

We have successfully built a Spring Boot application using GitHub Actions. The next section will use Skaffold and GitHub Actions to create a CI/CD pipeline for a Spring Boot application from the GitHub repository.

Creating a CI/CD pipeline with GitHub Actions and Skaffold

CI and CD are among the main pillars of the DevOps life cycle. As the name suggests, **continuous integration** (**CI**) is a software development practice where developers commit the code to a version control system several times a day. In **continuous deployment** (**CD**), software functionalities are delivered frequently through automated deployments, and there is no manual intervention or approval in this process. Only a failed test would halt your deployment to production. Another thing that is often confused with continuous deployment is continuous delivery, but they are different in reality. In continuous delivery, the main focus is on release and the release strategy and doing the actual deployment to production with approval. It is often termed **single-click deployment**.

By now, you will have developed some understanding of how GitHub Actions is event-driven and can automate your software development tasks. You will have also learned that you can trigger your entire CI/CD pipeline from your GitHub repository using GitHub Actions based upon certain events such as Git push or creating a pull request on a particular branch.

This section will focus on deploying a Spring Boot application to Google Kubernetes Engine using Skaffold and GitHub Actions. The workflow will closely mimic how we typically do the deployment in production using a CI/CD pipeline.

Before we proceed further with this task, there are a few prerequisites that we should be aware of. The following are some of the highlighted prerequisites.

Prerequisites

Please take a note of the following prerequisites:

- You need to create a new Google Cloud Project (or select an existing project). This part is already done in *Chapter 8, Deploying a Spring Boot Microservice to Google Cloud Platform Using Skaffold*, and we will use the same project.

- Please make sure that you enable the **Container Registry** and **Kubernetes Engine** APIs.

- You also have to create a new **Google Kubernetes Engine** (**GKE**) cluster or select an existing GKE cluster.

- If not done already, you also need to create a JSON service account key for the service account and add Kubernetes Engine Developer and Storage Admin roles. Service account keys are a safe way of accessing your cloud resources from outside. To establish the identity of a service account, a public/private key pair is used. The public key is stored in Google Cloud and the private key is available to you.

- To create a service account key, click on **IAM & Admin** on the left-hand side navigation bar on the Google Cloud Console. Click on **Service accounts** and then you will see the following screen:

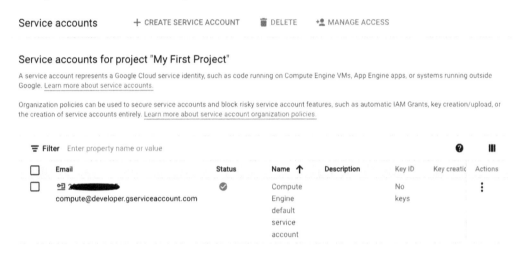

Figure 9.8 – Service account of your GCP project

- Now click the email address of the service account and select **Keys** from the right-hand side tab. Click on **ADD KEY** and select **Create new key,** as shown in the following screenshot:

Figure 9.9 – Adding a key to your service account

Choose **JSON** for **Key type** and click on **CREATE**. It will download the keys to your system, as shown here:

Create private key for "Compute Engine default service account"

Downloads a file that contains the private key. Store the file securely because this key can't be recovered if lost.

Key type

◉ JSON
Recommended

○ P12
For backward compatibility with code using the P12 format

CANCEL   CREATE

Figure 9.10 – Selecting the key type for your service account

- You need to add the following Cloud IAM roles to your service account:

a. **Kubernetes Engine Developer**: This role will allow you to deploy to GKE.

b. **Storage Admin**: This role will allow you to publish the container image to the Google Container registry:

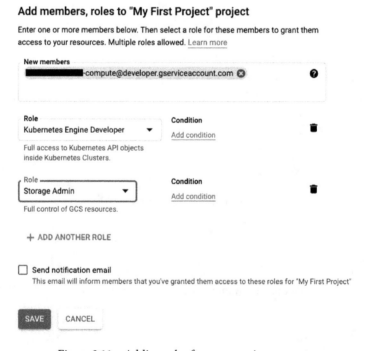

Figure 9.11 – Adding roles for your service account

- Add the following secrets to your GitHub repository's secrets. You can add GitHub repository secrets by navigating to the **Settings** tab and then clicking on **Secrets** on the left-hand side navigation bar. There, click on **New repository secret** and add the following secrets:

a. **PROJECT_ID**: The Google Cloud project ID

b. **SERVICE_ACCOUNT_KEY**: The content of the service account JSON file

Refer to the following screenshot:

Figure 9.12 – Adding secrets to your GitHub repository

With this, we have completed all the prerequisites. In the next section, we will create a CI/CD pipeline using GitHub Actions and Skaffold.

Implementing CI/CD workflow with GitHub Actions and Skaffold

In this section, we will create a production-ready CI/CD pipeline using Skaffold and GitHub Actions.

The following figure demonstrates the CI/CD workflow with Skaffold and GitHub Actions:

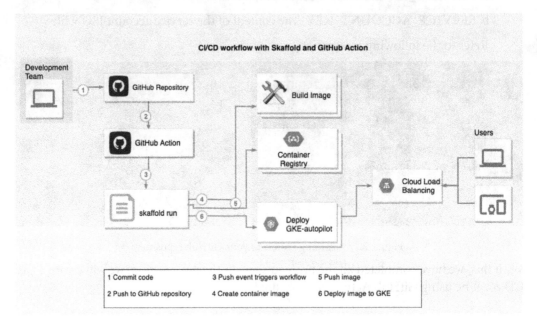

Figure 9.13 – CI/CD workflow with Skaffold

We will use the following workflow YAML file. Here, I have explained the workflow YAML file with comments in each step:

1. Specify the name and event of the workflow:

```
name: Deploy to GKE
on:
  push:
    branches:
      - main
```

2. Then we pass GitHub secrets as environment variable:

```
env:
    PROJECT_ID: ${{ secrets.PROJECT_ID }}
    GKE_CLUSTER: autopilot-cluster-1
```

```
    GKE_ZONE: us-central1
    SKAFFOLD_DEFAULT_REPO: gcr.io/${{ secrets.PROJECT_ID
    }}/breathe
```

3. Next, we define the job that runs on an Ubuntu Linux runner hosted by GitHub:

```
jobs:
  deploy:
    name: Deploy
    runs-on: ubuntu-latest
    env:
      ACTIONS_ALLOW_UNSECURE_COMMANDS: 'true'
```

4. While defining steps, the first step is to check out source code and then install Java 16:

```
    steps:
      - name: Check out repository on main branch
        uses: actions/checkout@v1
        with:
          ref: main
      - name: Install Java 16
        uses: AdoptOpenJDK/install-jdk@v1
        with:
          version: '16'
          architecture: x64
```

5. Then we set up the `gcloud` CLI:

```
      - name: Install gcloud
        uses: google-github-actions/setup-
            gcloud@master
        with:
          version: "309.0.0"
          service_account_key: ${{
            secrets.SERVICE_ACCOUNT_KEY }}
          project_id: ${{ secrets.PROJECT_ID }}
          export_default_credentials: true
```

6. Next, download `kubectl` for post-deployment verification and `skaffold` for continuous delivery:

    ```
    - name: Install kubectl and skaffold
      uses: daisaru11/setup-cd-tools@v1
      with:
        kubectl: "1.19.2"
        skaffold: "1.29.0"
    ```

7. Next, cache artifacts such as dependencies to improve workflow execution time:

    ```
    - name: Cache skaffold image builds & config
      uses: actions/cache@v2
      with:
        path: ~/.skaffold/
        key: fixed-${{ github.sha }}
    ```

8. Configure docker to use the gcloud command-line tool as a credential helper for authentication:

    ```
    - name: Configure docker
      run: |
        gcloud --quiet auth configure-docker
    ```

    Get the GKE credentials and deploy to the cluster using the `skaffold run` command, as follows:

    ```
    - name: Connect to cluster
      run: |
        gcloud container clusters get-credentials
          "$GKE_CLUSTER" --zone "$GKE_ZONE"
    ```

9. Finally, build and deploy to GKE using `skaffold run` and do verification with `kubectl get all` post deployment:

    ```
    - name: Build and then deploy to GKE cluster
            with Skaffold
      run: |
        skaffold run
    - name: Verify deployment
      run: |
        kubectl get all
    ```

You can use this workflow YAML file in your project and replace the secrets with your values. If you have placed the `skaffold.yaml` file in the root directory of your repository then it's OK, otherwise you can pass the `-filename` flag with the `skaffold run` command to point to the Skaffold configuration file.

If the workflow is executed successfully then you should see the following output:

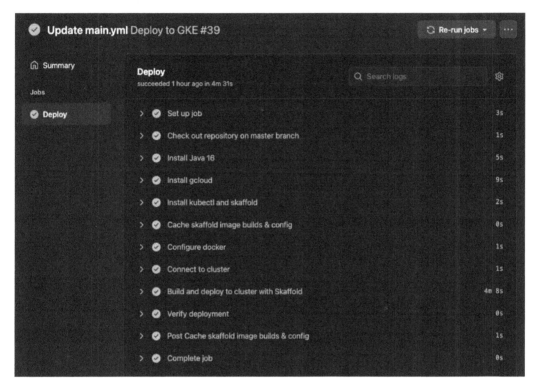

Figure 9.14 – Successful build and deployment to GKE with Skaffold

In this section, we have successfully built and deployed a Spring Boot application to a remote cluster with a customized CI/CD pipeline from the GitHub repository using Skaffold and GitHub Actions.

Next, let's see how we implement the workflow with Argo CD and Skaffold while understanding what they are.

Implementing a GitOps workflow with Argo CD and Skaffold

In *Chapter 4*, *Understanding Skaffold's Features and Architecture*, while explaining Skaffold features, we briefly talked about how we can use the `skaffold render` and `skaffold apply` commands to create a GitOps style continuous delivery workflow with Skaffold. In this section, we will implement a GitOps workflow using Skaffold and Argo CD. But first, let's understand what GitOps is and its benefit.

What is GitOps, and what are its benefits?

The word *GitOps* was coined by a company named Weaveworks. The idea behind GitOps is to consider Git as a single source of truth for your application and declarative infrastructure. Using Git to manage your declarative infrastructure makes it easy for developers because they interact with Git daily. Once you add configuration inside Git, you get the benefits of version control, such as reviewing changes using pull requests, audit, and compliance.

With GitOps, we create automated pipelines to roll out changes to your infrastructure when someone pushes changes to a Git repository. Then we use GitOps tools to compare the actual production state of your application with what you have defined under source control. Then it also tells you when your cluster doesn't match what you have in production and automatically or manually reconciles it with the desired state. This is true CD.

You can easily roll back your changes from Kubernetes by doing a simple `git revert`. In disaster scenarios or if someone accidentally nuked your entire Kubernetes cluster, we could quickly reproduce your whole cluster infrastructure from Git.

Now, let's understand a few benefits of GitOps:

- Using GitOps, the team is shipping 30-100 changes per day to production. Of course, you need to use deployment strategies such as blue-green and canary to validate your changes before making them available to all the users. The overall benefit is an increase in developer productivity.

- You get a better developer experience with GitOps as developers are pushing code and not containers. Moreover, they use familiar tools such as Git and don't need to know about the internals of Kubernetes (that is, `kubectl` commands).

- By putting declarative infrastructure as code in the Git repository, you automatically get benefits such as audit trail for your cluster, such as who did what and when. It further ensures the compliance and stability of your Kubernetes cluster.

- You can also recover your cluster faster, in case of a disaster, from hours to minutes because your entire system is described in Git.

- Your application code is already on Git, and with GitOps, your operation tasks are part of the same end-to-end workflows. You have a consistent Git workflow across your entire organization.

It's only fair that we also cover some details about Argo CD so that it's easier to understand the later part where we implement a GitOps workflow using Skaffold and Argo CD.

What is Argo CD?

As per the official documentation of **Argo CD**, `https://argo-cd.readthedocs.io/en/stable/`, it is a declarative, GitOps continuous delivery tool for Kubernetes. In the previous section, we used the term *GitOps tool* that can compare and sync the application state if it deviates from what we have defined in the Git repository, so it is safe to say that Argo CD is the tool that handles this automation. Kubernetes introduced us to the concept of control loops through which Kubernetes checks whether the number of replicas running matches with the desired number of replicas. Argo CD leverages the same **Kubernetes** (**K8s**) capabilities, and its core component is `argocd-application-controller`, which is basically a Kubernetes controller. It monitors the state of your application and adjusts the cluster accordingly.

And now it's time to learn about GitOps by implementing it with Skaffold and Argo CD on Google Kubernetes Engine. Let's begin.

Continuous delivery with Argo CD and Skaffold on GKE

Before we begin, we need to make sure that we have met the following prerequisites.

- We first need to install `kubectl`.
- The current Kubernetes context is set to a remote GKE cluster. You can verify the current context with the `kubectl config current-context` command.

We can run this demonstration on the local Kubernetes cluster but, ideally, you would be running it with a managed Kubernetes service such as GKE. Let's begin:

1.   First, we will install Argo CD on GKE using the following command:

```
kubectl create namespace argocd
kubectl apply -n argocd -f
https://raw.githubusercontent.com/argoproj/argo-cd/
stable/manifests/install.yaml
```

We created a separate namespace, `argocd`, and all Argo CD-related components will be part of it. We can verify the installation by navigating to the workloads section under GKE.

In the following screenshot, you can see that Argo CD stateful set components, that is, `argocd-application-controller`, and deployment components such as `argocd-server` are up and running on GKE:

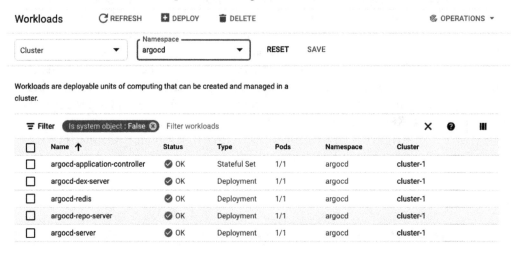

Figure 9.15 – Argo CD Kubernetes resources deployed to GKE

Next, we can install the Argo CD CLI. This is an optional step as we will be using the Argo CD UI instead.

2.   Next, we need to expose the Argo CD API server as by default it is not exposed for external access. We can run the following command to change the service type to `LoadBalancer`:

```
kubectl patch svc argocd-server -n argocd -p '{"spec":
{"type": "LoadBalancer"}}'
```

In the following screenshot, you can see that the service type has changed to `External load balancer` and we will access the Argo CD GUI using that IP address:

Figure 9.16 – Argo CD API server exposed as LoadBalancer

You can even use ingress or `kubectl` port forwarding for accessing the Argo CD API server without exposing the service.

3. We can now access the Argo CD GUI using the default admin user name and get the password using the following command:

```
kubectl -n argocd get secret argocd-initial-admin-secret
-o jsonpath="{.data.password}" | base64 -d
```

You should see the following login screen:

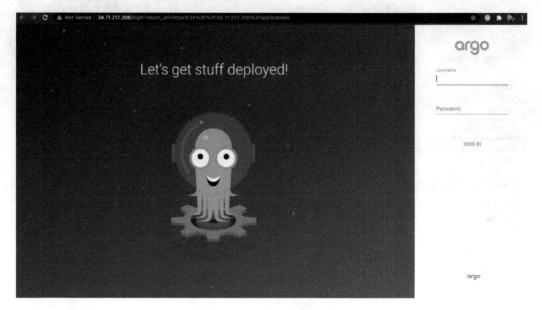

Figure 9.17 – Argo CD login screen

After logging in, click the **+ NEW APP** button, as shown in the following screenshot:

Figure 9.18 – Create application

In the next screen, enter your application name, choose the default project, and set **SYNC POLICY** to **Automatic**:

GENERAL

EDIT AS YAML

Application Name
air-quality-check

Project
default

SYNC POLICY
Automatic

☐ PRUNE RESOURCES
☐ SELF HEAL

SYNC OPTIONS

☐ SKIP SCHEMA VALIDATION   ☐ AUTO-CREATE NAMESPACE
☐ PRUNE LAST   ☐ APPLY OUT OF SYNC ONLY
☐ REPLACE ⚠

PRUNE PROPAGATION POLICY: foreground

Figure 9.19 – Argo CD application onboarding

4. Enter the source Git repository URL. Provide the path for Kubernetes manifests inside the Git repository. Argo CD polls your Git repository every 3 minutes to apply updated manifests to your Kubernetes cluster. You can avoid this delay by setting up a webhook event:

SOURCE

Repository URL
https://github.com/yrashish/Effortless-Cloud-Native-Apps-Development-using-Skaffold.git   GIT ▼

Revision
HEAD   Branches ▼

Path
Chapter09/gitops

Figure 9.20 – Providing application Git repository details to Argo CD

For **DESTINATION**, set cluster to in-cluster and **Namespace** to **default,** as shown in the following screenshot:

DESTINATION

Cluster URL
https://kubernetes.default.svc   URL ▾

Namespace
default

Figure 9.21 – Providing destination cluster details to Argo CD

After filling out the required information, click **CREATE** at the top of the UI to create the application. After clicking on the **CREATE** button, the Kubernetes manifests available in the Git repository at the path `Chapter09/gitops` are retrieved, and Argo CD performs `kubectl apply` on those manifests:

Figure 9.22 – Create application

After creating the application, you should see the following screen.
The **Status** is **Progressing**:

Figure 9.23 – Application created and synced

Click on the application and you will see the following screen:

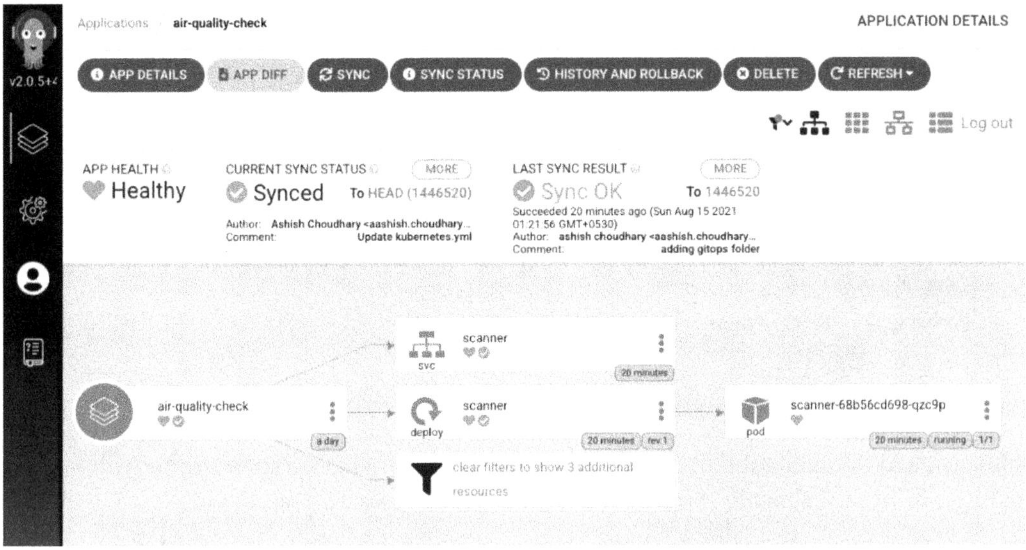

Figure 9.24 – Application deployed and in Healthy status

You can see the deployments, `svc`, and pods listed here. The application **SYNC STATUS** is **Synced** and **APP HEALTH** is **Healthy**. Argo CD has built-in health checks for different Kubernetes resource types such as Deployment and ReplicaSets.

We have set up the continuous delivery workflow for our application, and the application is synced successfully. Now we will try to test the workflow by doing some local changes with the following steps:

1. Set the default container registry to GCR using the `skaffold config set default-repo gcr.io/project-id` command.

2. We will build, tag, and push the container image using the `skaffold build` command.

3. Then we will run the `skaffold render` command. This command will generate hydrated (that is, with newly generated image tags) Kubernetes manifests to a file that we will later commit and push to the Git repository. The GitOps pipeline using Argo CD will pick and sync those changes to the target Kubernetes cluster.

Let's begin with this process.

We will make cosmetic code changes, increase replicas from one to two and run the `skaffold render` command. As per the `skaffold.yaml` file, the Kubernetes manifests are defined in the k8s directory. While running the `skaffold render` command, we will also pass the `--output=gtipos/manifest.yaml` flag so that we can later push it to the Git repository. The following is the output:

```
skaffold build && skaffold render --output=gtipos/manifest.yaml
Generating tags...
- breathe -> gcr.io/basic-curve-316617/breathe:99b8380-dirty
Checking cache...
- breathe: Not found. Building
Starting build...
Building [breathe]...
```

I just wanted to highlight that `skaffold render` doesn't generate anew but will use existing Kubernetes manifests and update the image with a new tag.

Finally, we commit the changes and push them to the GitHub repository with the following command:

```
git commit -m  "changing number of replicas" && git push
```

Soon after the push, Argo CD will sync the changes to the GKE cluster, as shown in the following screenshot:

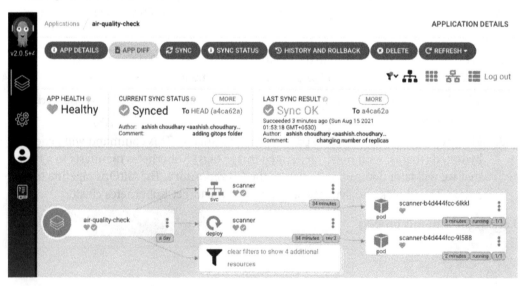

Figure 9.25 – Increased number of replicas

In the screenshot, you can see that now we have two pods running as we have increased the number of replicas.

The following screenshot illustrates a typical GitOps workflow with Skaffold and Argo CD. We have pretty much already covered the same steps so far. Let's try to summarize what we have learned so far.

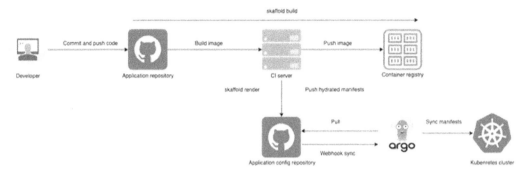

Figure 9.26 – GitOps workflow with Skaffold and Argo CD

We can conclude the following from the screenshot:

- A developer commits and pushes code changes to the Git repository.
- Continuous integration pipeline kicks in and using skaffold build, we will build, tag, and push the image to container registry.
- We will also generate hydrated manifests using skaffold render and commit them to either the same or different repository.
- Either sync action is triggered using CI webhook, or changes are pulled after regular polling intervals by Argo CD controller running inside the Kubernetes cluster.
- Further, the Argo CD controller will compare the live state against the desired target state (as per the git commit done on the config repository).
- If Argo CD detects that the application is OutOfSync, it will apply the latest changes to the Kubernetes cluster.

In this section, we have learned how to create a GitOps pipeline combining two powerful tools: Skaffold and Argo CD. We could have used the skaffold apply command instead of Argo CD but the skaffold apply command always uses kubetcl to deploy resources to the target cluster. If you have an application bundled as Helm charts then it will not work. Furthermore, with Argo CD, you can combine Argo Rollouts to do blue-green and canary deployments as they are natively not supported with Skaffold.

Summary

In this chapter, you have learned how you can use GitHub Actions to automate your development workflows. We started the chapter by explaining GitHub Actions and their components. We explained GitHub Actions and related concepts with an example. In the example, we explained how you could build, test, and deploy your Java applications from your GitHub repository. Then we described how you could create a CI/CD pipeline for your Kubernetes applications using Skaffold and GitHub Actions.

You have discovered how you can take advantage of GitHub Actions and combine them with Skaffold to create CI/CD pipelines. Then in the final section, we took deep dive into setting up GitOps style continuous delivery workflow with Skaffold and Argo CD. We have learned that in GitOps we consider Git repository as a single source of truth for any change related to your infrastructure. We have also demonstrated how we can implement GitOps pipeline with Argo CD and Skaffold.

In the next chapter, we will talk about Skaffold alternatives, and we will also cover its best practices and common pitfalls.

Further reading

- Learn more about automating workflow with GitHub actions from Automating Workflows with GitHub Actions (`https://www.packtpub.com/product/automating-workflows-with-github-actions/9781800560406`), published by Packt.

- Learn more about GitHub from GitHub Essentials (`https://www.packtpub.com/product/github-essentials-second-edition/9781789138337`), published by Packt.

10
Exploring Skaffold Alternatives, Best Practices, and Pitfalls

In the previous chapter, you learned how you can create a CI/CD pipeline for a Spring Boot application using GitHub Actions and Skaffold. In this chapter, we will start by looking at other tools available on the market that provide similar functionalities as Skaffold. We will learn about tips and tricks that developers can adhere to while developing cloud-native Kubernetes applications with Skaffold. We will conclude by understanding Skaffold pitfalls that developers can typically avoid.

In this chapter, we're going to cover the following main topics:

- Comparing Skaffold with other alternatives
- Applying Skaffold best practices
- Avoiding Skaffold pitfalls
- Future roadmap
- Final thoughts

By the end of this chapter, you will have a solid understanding of tools other than Skaffold for improving the developer experience with Kubernetes. You will also learn about Skaffold best practices that you can use for your development workflow and some common pitfalls that can be avoided in your development life cycle.

Comparing Skaffold with other alternatives

In this section, we will compare Skaffold with other alternative tools that address similar problems that Skaffold solves, that is, improving the developer experience with Kubernetes. However, there may be use cases where Skaffold may not be the best tool for the job, so we will look into tools other than Skaffold if your use case is complex in the next section. We will also look at the features that these Kubernetes development tools provide in comparison with Skaffold. Let's begin!

Telepresence

Telepresence (`https://www.telepresence.io/`) is a tool developed by Ambassador Labs. It's a Cloud Native Computing Foundation sandbox project. Its goal is to improve the developer experience with Kubernetes. Telepresence allows you to run a single service locally while connecting that service to a remote Kubernetes cluster. You can read more about it here: `https://www.telepresence.io/about/`.

With Telepresence, you can develop and debug a service locally as part of the cluster. You don't have to publish and deploy new artifacts in the Kubernetes cluster continuously. Telepresence doesn't require a local Kubernetes cluster in comparison with Skaffold. It runs a pod as a placeholder for your application in the remote cluster, and incoming traffic is routed to the container running on your local workstation. When a developer changes the application code, it will be reflected in your remote cluster without deploying a new container.

Another advantage is that you only need compute resources, namely, CPU and memory, to run your services locally as you are directly working with a remote cluster. Moreover, you don't have to set up and run a local Kubernetes cluster, such as minikube or Docker Desktop. It is helpful in cases where you have, say, five to six microservices running and your application must interact with them. In contrast, Skaffold is more of a complete solution packaged into one thing addressing your local development needs and CI/CD workflows. But let's suppose your application needs to interact with many microservices. In that case, it gets tricky as it would be difficult to run all the instances locally because of resource constraints, and you may end up mocking some of the services that may not replicate your actual production behavior. This is where Telepresence can help with its remote development capabilities from your laptop and with minimal resource usage.

There are some disadvantages, such as it has some known issues with volume mount for the Windows OS and needs a high-speed network connection.

Tilt.dev

Tilt (`https://tilt.dev/`) is an open source tool for improving the developer experience with Kubernetes.

In Skaffold, we use the `skaffold.yaml` configuration file to define, build, and deploy; and similarly, in Tilt, we use a Tiltfile for configuration. Tiltfiles are written in a dialect of Python called **Starlark**. Check out the API reference here: `https://docs.tilt.dev/api.html`. The following is a sample Tiltfile:

```
docker_build('example-html-image', '.')
k8s_yaml('kubernetes.yaml')
k8s_resource('example-html', port_forwards=8000)
```

Next, you can run the `tilt run` command. Tilt is not just a CLI tool in comparison with Skaffold. Tilt also provides a neat UI where you can view each of your service's health status, their build, and runtime logs. While Skaffold is an open source project with no vendor support, Tilt does offer vendor support with its enterprise edition.

The downside of Tilt is that you have to get familiar with its Starlark Python syntax language, which is required to write the Tiltfile, while Skaffold uses the `skaffold.yaml` configuration file as the YAML syntax file. But if you work with Kubernetes manifests, then this is not something new and most developers are already familiar with it.

OpenShift Do

Developing an application with Kubernetes or **Platform-as-a-Service** (**PaaS**) offerings such as OpenShift is hard if you are not using the correct tools throughout your development. In the case of OpenShift, it already has a CLI tool, **oc**, but unfortunately, it is more focused on helping operations folks and is not very developer-friendly. The oc CLI requires you to know and understand concepts related to OpenShift, such as deployment and build configurations to name a few, which, as a developer, you might not really be interested in knowing. The Red Hat team realized this issue and developed a new CLI called **OpenShift Do** (**odo**), which is more targeted toward developers. It also helps in improving the developer experience while developing cloud-native applications deployed to Kubernetes or OpenShift.

Let's look at some of its features, as follows:

- Develop faster and accelerate the inner development loop.

- Live feedback with the `odo watch` command. If you have worked with Skaffold, then it is quite similar to the `skaffold dev` command. The odo CLI uses developer-focused concepts such as **projects**, **applications**, and **components**.

- It's a completely CLI client-based tool.

OpenShift odo is very specific to OpenShift itself; even though the documentation says that it works with vanilla Kubernetes distributions. There is a lack of documentation and practical examples to use odo with minikube or other tools.

Oketo

Oketo (`https://okteto.com/`) is a CLI tool that takes an approach that is completely different than Skaffold. Instead of automating your inner development loop in your local workstation, Oketo moves the inner development loop to the cluster itself. You define your development environment in a YAML manifest file, `okteto.yaml`, and then use `okteto init` and `okteto up` to get up and running with your development environment.

The following are some of the highlighted features of Oketo:

- File synchronization between the local and remote Kubernetes cluster.

- A single binary that works across different OSes.

- You get a remote terminal in your container development environment.

- Hot reload your code.

- Works with local and remote Kubernetes clusters, Helm, and serverless functions.

- Port forwarding in both directions.

- No building, pushing, or deploying is required as you develop directly on your cluster.

- No need to install Docker or Kubernetes on your workstation.

- Even runtimes (JRE, npm, Python) are not required as everything is inside a Docker image.

Garden

Garden (`https://garden.io/`) follows the same philosophy as Oketo of deploying to a remote cluster instead of doing the setup on your local system. Garden is an open source tool that runs Kubernetes applications in a remote cluster for development, automated testing, manual testing, and review.

With Garden, you can start by using CLI helper commands such as `garden create project`. You will manage the Garden configuration through a YAML configuration file.

The following are the key elements of the Garden YAML configuration file:

- **Modules**: In modules, you specify how to build your containers.

- **Services**: In services, you specify how to run your containers on Kubernetes.

- **Tests**: In tests, you specify unit and integration tests.

The following are some of the features that Garden provides:

- It automatically redeploys the application to the remote cluster when the source code is changed.

- It supports multi-module and multi-service operations (tree of dependencies).

- It provides a graphical dashboard for dependencies.

- The ability to run tasks (for example, database migrations as part of the build flow).

- It supports Helm and OpenFass deployments.

- It supports hot reload features where source code is sent directly to running containers.

- The ability to stream container logs to your terminal.

- It supports file watching and hot reloading of code for remote clusters.

Garden has a more complex setup than Skaffold, and it takes a while to get familiar with its concepts, so there is a steep learning curve involved. With Skaffold, you work with familiar build and deploy tools, and it's easy to get started with it. It might also be overkill to use Garden for small applications due to its inherent complexities. Garden is commercial open source, so some of its features are paid compared to Skaffold, which is entirely open source.

docker-compose

docker-compose is a tool that is primarily used for local development with containers. It allows you to run multiple containers locally and mimic how an application would look when deployed to Kubernetes. Docker needs to be installed on the workstation to get started with it. While docker-compose may give some developers a false impression of running their application in a Kubernetes environment such as minikube, in reality, it is nothing like running it in a Kubernetes cluster. It also means that because your application works on docker-compose, it will not work or behave similarly when deployed to the Kubernetes cluster in production. While we know that containers solve the problem of *works on my machine*, with docker-compose, we introduce a new problem, that is, *works on my docker-compose setup*.

It may be tempting to use docker-compose as a replacement to ease the inner development loop of cloud applications, but as explained earlier, your local and production environment will not be the same. It would be hard to debug any environment because of this difference, while with Skaffold, you get to use the exact same stack for your local and remote build and deployment. If, for some reason, you are stuck with the docker-compose setup, you can even pair it with Skaffold. Skaffold internally uses Kompose (https://kompose.io/) to convert docker-compose.yaml into Kubernetes manifests.

You can use the following command to use your existing docker-compose.yaml file with Skaffold:

```
skaffold init --compose-file docker-compose.yml
```

In this section, we have looked at Kubernetes development tools other than Skaffold, helping developers develop faster and get quick feedback in their inner development loop.

In the next section, we will learn about some best practices to apply to our existing or new workflow with Skaffold.

Applying Skaffold best practices

In this section, we will learn about Skaffold best practices that you, as a developer, can take advantage of, to either speed up your deployment in the inner or outer development loop or use some flags to make things easier while using Skaffold. Let's begin:

- While working with multiple microservices applications deployed to Kubernetes, sometimes, it's challenging to create a single `skaffold.yaml` configuration file for each application. In those common cases, you can create `skaffold.yaml` scoped for each application, and then run the `skaffold dev` or `run` command independently for each application. You can even iterate both the applications together in a single Skaffold session. Let's assume we have a frontend app and a backend app for both of them; your single `skaffold.yaml` file should look like the following:

```
apiVersion: skaffold/v2beta18
kind: Config
requires:
- path: ./front-end-app
- path: ./backend-app
```

When you are bootstrapping your project with Skaffold and you don't have Kubernetes manifests, you can pass the `--generate-manifests` flag with the `skaffold init` command to generate basic Kubernetes manifests for your project.

- It would be best if you always use the `default-repo` functionality with Skaffold. If you are using `default-repo`, you don't have to manually edit the YAML files as Skaffold can prefix the image names with the container image registry specified by you. So, instead of writing `gcr.io/myproject/imagename`, you can enter the image name in the `skaffold.yaml` configuration file. Another advantage is that you can share your `skaffold.yaml` file easily with other teams as they don't have to manually edit the YAML file if they use a different container image registry. So, basically, you don't have to hardcode the container image registry names inside your `skaffold.yaml` configuration files by using the `default-repo` functionality.

You can utilize the `default-repo` functionality in the following three ways:

a. By passing the `--default-repo` flag:

```
skaffold dev --default-repo gcr.io/myproject/imagename
```

b. By passing the `SKAFFOLD_DEFAULT_REPO` environment variable:

```
SKAFFOLD_DEFAULT_REPO= gcr.io/myproject/imagename
skaffold dev
```

c. By setting Skaffold's global config:

```
skaffold config set default-repo gcr.io/myproject
/imagename
```

- It gets tricky to know the actual issue when you run into an issue with the Skaffold command. In some cases, you may need more information than what Skaffold typically displays while streaming logs. For such cases, you can use the -v or -verbosity flag to use a specific log level. For example, you use skaffold dev -v info to view information-level logs.

  Skaffold supports the following log levels, and the default is warn:

  - info
  - warn
  - error
  - fatal
  - debug
  - trace

- For a faster build, you can take advantage of the concurrency flag by setting it to 0. The default value is 0, meaning no limits on the number of parallel builds so all your builds are done in parallel. Only in the case of local build concurrency will the value default to 1, which means the build will be done sequentially to avoid any side effects.

- If you are using Jib to build and push container images, then you can use the special sync support using the auto configuration. You can enable it by using the sync: option, as mentioned in the following skaffold.yaml file:

```yaml
apiVersion: skaffold/v2beta16
kind: Config
build:
  artifacts:
    -
      image: file-sync
      jib: {}
      sync:
        auto: true
deploy:
  kubectl:
    manifests:
      - k8s-*
```

With this option, Jib can sync your class files, resource files, and Jib's *extra directories* files to a container running locally or remotely. You don't have to rebuild, redeploy, or restart the pod for each change in your inner development loop. However, for this to work with your Spring Boot application, you need to have the `spring-boot-devtools` dependency in your `pom.xml` file for your Maven project.

- Skaffold also supports Cloud Native Buildpacks to build your container images and, similar to Jib, it also supports the `sync` option to automatically rebuild and relaunch your application when changes are made to a certain type of file. It supports the following type of source files:

  - **Go**: `*.go`

  - **Java**: `*.java`, `*.kt`, `*.scala`, `*.groovy`, `*.clj`

  - **Node.js**: `*.js`, `*.mjs`, `*.coffee`, `*.litcoffee`, `*.json`

In this section, we have learned some best practices that we can apply to develop efficiently and accelerate the development loop even faster with Skaffold.

In the next section, we will be looking at some of the common Skaffold traps that we, as developers, should be aware of.

Avoiding Skaffold pitfalls

Throughout this book, we have used various features provided by Skaffold. Now let's discuss some common Skaffold pitfalls that we, as developers, should understand and try to avoid:

- Skaffold requires that either you have some local or remote Kubernetes setup so in comparison with other tools that we discussed in the previous section Skaffold doesn't reduce the time required to set up your development environment.

- With Skaffold, in most cases, you work with some local Kubernetes such as minikube or Docker Desktop and you cannot replicate your entire production-like setup with them because of their limitations. This leaves space for integration issues that you may not see on local systems but could pop up in higher environments.

- Sometimes, more hardware resources are wasted on your machine with Skaffold. For example, if you need to run, let's say, 10 microservices, then it becomes challenging as you're limited by resources on your laptop.

- Skaffold has built-in support for integrating with debuggers via the (beta) `skaffold debug` command. With this debugging option, Skaffold automatically configures the application runtime for remote debugging. It is a great feature, but using debuggers in a microservices environment is tricky at best. It gets even tougher working with remote clusters. Use it judiciously.

- Skaffold has no web UI. While we discussed many tools in the previous section that provide a UI for a better experience, I would not cry about it. It is more of a personal preference as some people tend to prefer a UI and some a CLI. If you are more of a UI person, then you may not get along with Skaffold.

- Skaffold is excellent for local development and testing in your inner development loop. Even though it is marketed as a complete CI/CD solution for some use cases, it may not be the best tool for the job. For example, if we want to scale to production or pre-production use cases, it is better to use dedicated tools for that such as **Spinnaker pipelines** (https://spinnaker.io/) or **Argo Rollouts** (https://argoproj.github.io/argo-rollouts/). These tools Spinnaker/Agro Rollouts provide some advantages over Skaffold. Let's see them:

  I.   In the case of Spinnaker/Agro Rollouts, both can support sophisticated/complex deployment strategies. You can define deployment strategies such as canary and blue/green deployment, stuff like that.

  II. Spinnaker allows multi-cluster deployments. Also, you can configure easy UI-based deployment to multiple clusters.

  III. Spinnaker has great visualization. It provides a rich UI that displays any deployment or pod status across clusters, regions, namespace, and cloud providers.

In this section, we have covered Skaffold pitfalls that you should look out for before deciding to go ahead with it for your Kubernetes workloads.

Future roadmap

The community primarily drives the Skaffold roadmap since it is an open source tool, and the team of engineers from Google makes the final call. Google developers also propose exciting new features that would enhance the user experience with Skaffold on top of the changes requested by the community.

However, a roadmap should not be considered a list of promises delivered no matter what. It is a sort of wish list that the Skaffold engineering team thinks could be worth investing their time on. The primary motivation behind the roadmap is to get feedback from the community around the features they want to see in Skaffold.

You can view the Skaffold roadmap for 2021 by accessing the `https://github.com/GoogleContainerTools/skaffold/blob/master/ROADMAP.md#2021-roadmap` URL.

Final thoughts

Tooling around Kubernetes developer tools has improved significantly in recent years. The primary motivation for that is the increased adoption of Kubernetes in the industry. Modern age developers want a tool that increases their productivity while developing applications for the cloud. Skaffold dramatically enhances the productivity of developers building and deploying Kubernetes applications.

Many tools internally use Skaffold, such as Jenkins X and Cloud Code, to improve the overall developer experience with Kubernetes. In contrast to Jenkins X, which uses Skaffold to build and push the image in the pipeline, Cloud Code is entirely built around Skaffold and its supported tools to provide a seamless onboarding experience for Kubernetes applications.

Finally, I would like to conclude by saying that Skaffold simplifies Kubernetes development, and in my opinion, it is doing a good job. It provides flexibility and extensibility on what kind of integrations it should be used with. Its extensible and pluggable architecture allows developers to choose an appropriate tool for each step involved in building and deploying the application.

Summary

In this chapter, we started by comparing Skaffold with other tools such as Tilt, Telepresence, Garden, Oketo, `docker-compose`, and OpenShift odo. These tools, in principle, try to provide a solution to a similar problem that Skaffold addresses. Then, we covered features that these tools offer in comparison to Skaffold. We also looked at some best practices that we can use with Skaffold for a better developer experience. Finally, we concluded by explaining some pitfalls related to Skaffold, which you should watch out for if your use case is more advanced.

You have discovered how you can take advantage of Skaffold by following some of the best practices that we have tried to explain. You are now in a better position to decide whether Skaffold fulfills your use case, or whether you need to consider other options we have covered in this chapter.

With this, we have reached the end of the journey and I hope you are encouraged to try out and explore a lot of Skaffold!

`Packt.com`

Subscribe to our online digital library for full access to over 7,000 books and videos, as well as industry leading tools to help you plan your personal development and advance your career. For more information, please visit our website.

Why subscribe?

- Spend less time learning and more time coding with practical eBooks and Videos from over 4,000 industry professionals

- Improve your learning with Skill Plans built especially for you

- Get a free eBook or video every month

- Fully searchable for easy access to vital information

- Copy and paste, print, and bookmark content

Did you know that Packt offers eBook versions of every book published, with PDF and ePub files available? You can upgrade to the eBook version at `packt.com` and as a print book customer, you are entitled to a discount on the eBook copy. Get in touch with us at `customercare@packtpub.com` for more details.

At `www.packt.com`, you can also read a collection of free technical articles, sign up for a range of free newsletters, and receive exclusive discounts and offers on Packt books and eBooks.

Other Books You May Enjoy

If you enjoyed this book, you may be interested in these other books by Packt:

Cloud Native with Kubernetes

Alexander Raul

ISBN: 9781838823078

- Set up Kubernetes and configure its authentication
- Deploy your applications to Kubernetes
- Configure and provide storage to Kubernetes applications
- Expose Kubernetes applications outside the cluster
- Control where and how applications are run on Kubernetes
- Set up observability for Kubernetes
- Build a continuous integration and continuous deployment (CI/CD) pipeline for Kubernetes
- Extend Kubernetes with service meshes, serverless, and more

Mastering Kubernetes - Third Edition

Gigi Sayfan

ISBN: 9781839211256

- Master the fundamentals of Kubernetes architecture and design
- Build and run stateful applications and complex microservices on Kubernetes
- Use tools like Kubectl, secrets, and Helm to manage resources and storage
- Master Kubernetes Networking with load balancing options like Ingress
- Achieve high-availability Kubernetes clusters
- Improve Kubernetes observability with tools like Prometheus, Grafana, and Jaeger
- Extend Kubernetes working with Kubernetes API, plugins, and webhooks

If you enjoyed this book, you may be interested in these other books by Packt:

Kubernetes in Production Best Practices

Aly Saleh, Murat Karslioglu

ISBN: 9781800202450

- Explore different infrastructure architectures for Kubernetes deployment
- Implement optimal open source and commercial storage management solutions
- Apply best practices for provisioning and configuring Kubernetes clusters, including infrastructure as code (IaC) and configuration as code (CAC)
- Configure the cluster networking plugin and core networking components to get the best out of them
- Secure your Kubernetes environment using the latest tools and best practices
- Deploy core observability stacks, such as monitoring and logging, to fine-tune your infrastructure

Packt is searching for authors like you

If you're interested in becoming an author for Packt, please visit authors. packtpub.com and apply today. We have worked with thousands of developers and tech professionals, just like you, to help them share their insight with the global tech community. You can make a general application, apply for a specific hot topic that we are recruiting an author for, or submit your own idea.

Share Your Thoughts

Now you've finished *Effortless Cloud-Native App Development Using Skaffold*, we'd love to hear your thoughts! Scan the QR code below to go straight to the Amazon review page for this book and share your feedback or leave a review on the site that you purchased it from.

https://packt.link/r/1801077118

Your review is important to us and the tech community and will help us make sure we're delivering excellent quality content.

Index